T0135470

Time-Periodic Solutions to the Equations of Magnetohydrodynamics with Background Magnetic Field

Vom Fachbereich Mathematik

der Technischen Universität Darmstadt

zur Erlangung des Grades eines

Doktors der Naturwissenschaften

(Dr. rer. nat.)

genehmigte

Dissertation

von

Jens-Henning Möller

aus Braunschweig

Referent:	Prof. Dr. Reinhard Farwig
Korreferent:	Prof. Dr. Mads Kyed
Tag der Einreichung:	09. Juli 2020
Tag der mündlichen Prüfung:	14. August 2020

Darmstadt 2020

Bibliographic information published by the Deutsche Nationalbibliothek

The Deutsche Nationalbibliothek lists this publication in the Deutsche
Nationalbibliografie; detailed bibliographic data are available
on the Internet at http://dnb.d-nb.de .

zugl.: Darmstadt, Technische Universität Darmstadt, Dissertation - D17
URN urn:nbn:de:tuda-tuprints-140886
URI https://tuprints.ulb.tu-darmstadt.de/id/eprint/14088

ISBN 978-3-8325-5187-2

Logos Verlag Berlin GmbH
Georg-Knorr-Str. 4, Gebäude 10
D-12681 Berlin
Tel.: +49 (0)30 42 85 10 90
Fax: +49 (0)30 42 85 10 92
https://www.logos-verlag.de

Contents

Introduction

The equations of magnetohydrodynamics, often abbreviated as MHD equations, describe the motion of an electrically neutral conducting fluid like plasmas, highly concentrated salt water and liquid metals. When considering this type of fluids, the classical Navier-Stokes equations do not adequately describe their motion, since they do not account for the additional electromagnetic forces influencing the motion, in this case given by the Lorentz force. We will give a brief derivation of the equations following Jackson [49, Chapter 10], where we skip physical arguments and at the moment do ignore any physical constants that might occur. For a more detailed approach we refer to Davidson [25] or Landau and Lifshitz [63, Chapter VIII], where the underlying physical considerations are presented in an easily comprehensible manner. The behaviour of a viscous incompressible fluid is described by the velocity $u(t,x)$ and the pressure $\mathrm{p}(t,x)$ for time t in some interval $(0,T)$ and point $x \in \Omega$ in some spatial domain $\Omega \subset \mathbb{R}^3$. The momentum equation is given by

$$\partial_t u - \Delta u + \nabla \mathrm{p} + (u \cdot \nabla)u = F + (J \times B) \quad \text{in } (0,T) \times \Omega$$

with an external force F acting on the fluid, *e.g.*, the gravitational force, current density J and magnetic field B. We are considering incompressible fluids, hence the continuity equation is given by $\operatorname{div} u = 0$. Therefore, the force equations of the Navier-Stokes and MHD equations seem quite similar, but because the Lorentz force $J \times B$ is influenced by the motion of the fluid we need to consider additional equations. Since it is an electromagnetic force, we have to consider the Maxwell equations. For magnetic field E and charge density ρ they are given by

$$(\text{ME}) \qquad \begin{cases} \operatorname{div} E = \rho, \\ \nabla \times B - \partial_t E = J, \\ \nabla \times E + \partial_t B = 0, \\ \operatorname{div} B = 0, \end{cases}$$

where as before we omitted constants. This formulation is also applicable for considerations on a microscopic level. For the MHD equations we neglect displacement current and hence the second equations simplifies to $\nabla \times B = J$. From physical considerations, *i.e.*, Ohm's law, we derive $J = E + u \times B$. To obtain the time evolution of B within the equations and to eliminate E, we apply the rotation to the previous identities and combine them with the momentum equations to conclude

$$\begin{cases} \partial_t u - \Delta u + \nabla \mathrm{p} + (u \cdot \nabla) u = F + ([\nabla \times B] \times B) & \text{in } (0, T) \times \Omega, \\ \nabla \times [\nabla \times B] = -\partial_t B + \nabla \times [u \times B] & \text{in } (0, T) \times \Omega, \\ \operatorname{div} B = 0, \ \operatorname{div} u = 0 & \text{in } (0, T) \times \Omega. \end{cases}$$

To examine these equations from a mathematical point of view, one needs to supplement the system with boundary conditions. The standard boundary conditions for these equations are a no-slip boundary condition for u, *i.e.*, $u = 0$ on $\partial \Omega$, and the perfect conductivity condition for E, *i.e.*, $E \times n = 0$ on $\partial \Omega$ where n is the outer normal vector to Ω. We briefly follow the ideas of Yoshida and Giga [87] to derive boundary conditions for B. Since $\nabla \times B = J = E + u \times B$, we obtain $[\nabla \times B] \times n = 0$ on $\partial \Omega$ because $E \times n = 0$ and u vanishes on the boundary. By making use of the divergence theorem we see that $E \times n = 0$ implies $[\nabla \times E] \cdot n = 0$ and thus we obtain $(\partial_t B) \cdot n = 0$ by the third equation of (ME). This implies $B \cdot n = B_1$ for given B_1. Before we state the full system of equations, we remark that the magnetic field B will often also be denoted by H. From a mathematical point of view the notations are interchangeable, but in physics they refer to different but very closely related fields. In a vacuum they differ by a factor, but otherwise by an additional term M called the magnetization of the material, for details see Jackson [49, Section 5.8]. For this thesis we will denote the magnetic field by H. Since $\operatorname{div} H = 0$ it holds $\nabla \times [\nabla \times H] = -\Delta H$ and together with $[\nabla \times H] \times H = (H \cdot \nabla) H - \frac{1}{2} \nabla |H|^2$, the non-dimensional equations of magnetohydrodynamics are given by

(MHDT) $\quad \begin{cases} \partial_t u - \nu \Delta u + \nabla \mathrm{p} + \dfrac{S}{2} \nabla |H|^2 + (u \cdot \nabla) u = F + S \cdot (H \cdot \nabla) H & \text{in } (0, T) \times \Omega, \\ \partial_t H - \mu \Delta H = \nabla \times [u \times H] & \text{in } (0, T) \times \Omega, \\ \operatorname{div} u = 0, \ \operatorname{div} H = 0 & \text{in } (0, T) \times \Omega, \\ u = 0, \quad H \cdot n = B_1, \quad \operatorname{curl} H \times n = 0 & \text{on } (0, T) \times \partial \Omega, \end{cases}$

with constants μ, ν, S. For details on the constants we refer to Landau and Lifshitz [63, Chapter VIII] or Sermange and Temam [75]. The functions B_1 and F are given and since the time-derivative of $H \cdot n$ vanishes on the boundary by the previous considerations, we derive that B_1 does not depend on time. This means that the magnetic field B_1 can be seen as the intrinsic field of the medium of the boundary.

Before we continue, we briefly remark some of the existing theory for the initial value problem of (MHDT), *i.e.*, under the additional conditions $u(0, x) = u_0(x)$ and $H(0, x) = H_0(x)$ for known functions u_0 and H_0. Existence of weak and strong solutions in L_2 can be found in Ladyženskaya and Solonnikov [62, Chapter 6], Giga and Yoshida [87] and Sermange and Temam [75]. The last paper collects a variety of general existence and uniqueness results in its Chapter 3 and only Giga and Yoshida considered $B_1 \neq 0$. Note that for existence of strong solutions some kind of smallness has to be assumed, either in the time $T > 0$ or in the norms of the initial values (u_0, B_0) and the external force F.

For results in L_p we note that global existence of solutions has been proven in L_3 by Akiyama [2] in a bounded domain and by Yamaguchi [84] in an exterior domain. Li and Wang [64] considered the problem in $L_p(L_q)$ for a bounded domain and showed local and global existence of solutions for $\frac{p}{2}(1 - \frac{3}{q}) \in (0, 1)$. Note that their result does not include the solutions constructed by Akiyama. The focus of this thesis lies on the existence of time-periodic solution to (MHDT), *i.e.*, we want to find solutions (u, H, p) such that $u(t + \mathcal{T}, x) = u(t, x)$, $H(t + \mathcal{T}, x) = H(t, x)$ and $\mathrm{p}(t + \mathcal{T}, x) = \mathrm{p}(t, x)$

for all $t \in \mathbb{R}$ and a given time-period $\mathcal{T} > 0$. One of the most popular ways to obtain a time-periodic solution is to employ the so-called Poincaré operator. The idea of this operator is to map the initial value to the value of the corresponding solution at time $\mathcal{T}$. For the MHD equations this would mean one maps (u_0, H_0) onto $(u(\mathcal{T}), H(\mathcal{T}))$ with (u, H) solving (MHDT). A periodic solution would exist if this mapping has a fixed point. The main challenge for this approach is to find Banach spaces such that this operator is well-defined and a fixed point exists. An early application can be found in Browder [19] for example.

A different approach is due to Arendt and Bu [9], who showed that a good understanding of an operator, e.g., the generation of a C_0-semigroup, maximal regularity of the initial value problem and invertibility, suffices for the existence of time-periodic solutions to the corresponding linear problem. Additionally, they showed that existence of periodic solution can also be concluded from $\mathcal{R}$-boundedness of a sequence of resolvents. An application of their result to time-periodic problems can for example be found in Hieber and Stinner [45].

All of the previous approaches conclude existence of time-periodic solutions from the study of the associated initial value problem. Recently Kyed [60] proposed a different way to approach time-periodic problems. He reformulated the problem into a time-periodic setting by considering the problem not on $\mathbb{R}$ but on $\mathbb{T} = \mathbb{R}/\mathcal{T}\mathbb{Z}$ and by decomposing the problem into a *steady-state* and a *purely periodic* part. The decomposition idea can also be found in Galdi [36, 37] and was developed in a joint effort. Before we explain the advantage of their approach, we remark that the reformulation onto $\mathbb{T}$ has been used before, see for example Rabinowitz [69, 70]. But the approaches were limited to L_2, since they relied on the usage of Plancherel's theorem, or to the space of absolutely convergent Fourier series. Therefore, the central idea of their approach is the splitting of the problem and considering the resulting problem separately in different function spaces.

To obtain the decomposition, for a time-periodic f defined on $(0, \mathcal{T}) \times \Omega$ we introduce the projections

$$\mathcal{P}f := \frac{1}{\mathcal{T}} \int_0^{\mathcal{T}} f(s, x) \, \mathrm{d}t, \quad \mathcal{P}_\perp f := f - \mathcal{P}f.$$

Since the function $\mathcal{P}f$ is independent of time, it is the mentioned steady-state, and $\mathcal{P}_\perp f$ is the purely periodic part. The stationary part of the linearised problem can usually be handled by standard theory for elliptic equations. For the purely periodic part the idea is to use the Fourier transform $\mathscr{F}_{G_n}$ on $G_n := \mathbb{T} \times \mathbb{R}^n$ given by

(FT) $$\mathscr{F}_{G_n}[f](k, \xi) := \frac{1}{\mathcal{T}(2\pi)^{\frac{n}{2}}} \int_0^{\mathcal{T}} \int_{\mathbb{R}^n} f(t, x) \, \mathrm{e}^{-ix \cdot \xi - i\frac{2\pi}{\mathcal{T}}kt} \, \mathrm{d}x\mathrm{d}t.$$

The resulting function $\mathscr{F}_{G_n}[f]$ is defined on $\mathbb{Z} \times \mathbb{R}^n$, and the purely periodic part satisfies

$$\mathscr{F}_{G_n}[\mathcal{P}_\perp f](0, \xi) = 0.$$

To construct time-periodic functions, a combination of classical Fourier multiplier results and a transference principle can be applied to yield existence of solutions on $\mathbb{T} \times \mathbb{R}^n$. Afterwards, classical methods of reflection and localisation can be used to construct solutions in sufficiently smooth domains. Applications of this technique can for example be found in Celik and Kyed [20, 21], Eiter and Kyed [30] or Kyed and Sauer [61].

The advantage of this approach is clear: One directly constructs time-periodic solutions and therefore avoids considerations of initial value problems or the concept of $\mathcal{R}$-boundedness. But since this strategy is rather new, the general theory of time-periodic function spaces is not as developed as for the initial value problems. A part of this thesis aims the development of further theory on time-periodic function spaces. For Lebesgue and Sobolev spaces it is readily checked that most of the properties needed for the theory of parabolic boundary problems hold true in the time-periodic framework as well. Additional theory is needed when, similarly to the work of Li and Wang [64], considering the problem in $L_q(\mathbb{T}, L_p(\Omega))$. It is known that in this setting the problem of boundary traces becomes more involved, see Weidemaier [81] and Denk, Hieber and Prüss [27]. Namely the resulting trace space is an (anisotropic) Triebel-Lizorkin space. The cited works determine it as an intersection of Triebel-Lizorkin spaces, whereas later work of Johnsen and Sickel [56] proved the anisotropic result. Since these trace results are fundamental for a comprehensive theory of time-periodic boundary value problems, we will extend the results of Johnsen and Sickel in Chapter 3 to time-periodic functions.

For classical Triebel-Lizorkin spaces one takes a partition of unity of $\mathbb{R}^n$ with smooth functions $(\varphi_j)_{j \in \mathbb{N}_0}$ such that their supports are contained in some annuli $A_j := \{x \in \mathbb{R}^n \mid A2^{j-1} \leq |x| \leq B2^j\}$ for $j \in \mathbb{N}$ and suitable constants A, B, and a ball A_0 containing the origin such that $\sum_{j=0}^{\infty} \varphi_j(x) = 1$ for all $x \in \mathbb{R}^n$. For the anisotropic version the idea is similar, but makes use of an anisotropic distance function introduced by Fabes and Rivière [32] and extended by Yamazaki [85], see Section 1.4. It results in a decomposition where the support of the functions φ_j is not rotational symmetric with respect to its arguments but distorted by a vector $\vec{a}$. The norm is defined in the same way. For $u \in \mathscr{S}'(\mathbb{R}^n)$ set

$$u_j := \mathscr{F}_{\mathbb{R}^n}^{-1} \varphi_j \mathscr{F}_{\mathbb{R}^n} u \quad \text{so that} \quad u = \sum_{j=0}^{\infty} u_j$$

and define the norm of anisotropic Triebel-Lizorkin spaces $\mathrm{F}_{\vec{p},r}^{s,\vec{a}}(\mathbb{R}^n)$ with regularity parameter $s \in \mathbb{R}$ by

$$\|u\|_{\mathrm{F}_{\vec{p},r}^{s,\vec{a}}(\mathbb{R}^n)} := \left\| \{2^{sj} u_j\}_{j \in \mathbb{N}_0} \right\|_{L_{\vec{p}}(\mathbb{R}^n, \ell_r(\mathbb{N}_0))}.$$

The $L_{\vec{p}}$-norm is applied in each variable inductively, *i.e.*, the $L_{p_i}(\mathbb{R})$-norm is applied with respect to the variable x_i. Additionally, a scalar in the place of a vector denotes the vector that is constant in its entries, *e.g.*, $\vec{a} = 1$ means $\vec{a} = (1, 1, \ldots, 1)$. Standard theory shows $\mathrm{F}_{\vec{p},2}^{k,1}(\mathbb{R}^n) = \mathrm{W}_{\vec{p}}^k(\mathbb{R}^n)$ for $k \in \mathbb{N}_0$, and it holds

$$\mathrm{F}_{(q,p),2}^{2,(2,1)}(\mathbb{R} \times \mathbb{R}^n) = \mathrm{W}_q^1(\mathbb{R}, L_p(\mathbb{R}^n)) \cap L_q(\mathbb{R}, \mathrm{W}_p^2(\mathbb{R}^n)),$$

which is the classical parabolic space. The last identity is one of the major reasons why these spaces are of interest.

In Section 3.1 we define the corresponding time-periodic version $\mathrm{F}_{(q,\vec{p}),r}^{s,(b,\vec{a})}(\mathbb{T} \times \mathbb{R}^n)$ with anisotropy $(b, \vec{a})$ and show well-definedness and properties similar to those of classical spaces. These spaces have not been considered beforehand and some of the classical results and approaches are not applicable due to the structure of the underlying space $\mathbb{T} \times \mathbb{R}^n$. Therefore, we prove a transference principal for vector-valued multipliers and an anisotropic Littlewood-Paley decomposition in Sections 2.2 and 2.3, which in turn can be used to reproduce the identity

$$\mathrm{W}_q^1(\mathbb{T}, L_p(\mathbb{R}^n)) \cap L_q(\mathbb{T}, \mathrm{W}_p^2(\mathbb{R}^n)) = \mathrm{F}_{(q,p),2}^{2,(2,1)}(\mathbb{T} \times \mathbb{R}^n).$$

As a next step we come back to the trace problem and we see the advantage of working with Triebel-Lizorkin spaces in this context. By the Paley-Wiener-Schwartz theorem it is well-known that the inverse Fourier transform of a compactly supported distribution in $\mathbb{R}^n$ is a smooth function, hence $u_j = \mathscr{F}_{\mathbb{R}^n}^{-1} \varphi_j \mathscr{F}_{\mathbb{R}^n} \in C^\infty(\mathbb{R}^n)$ and therefore the trace $u_j|_{\mathbb{R}^{n-1}}$ is a well-defined function. Thus a working definition of a trace of an element of a Triebel-Lizorkin spaces is

$$T_n(u) := \sum_{j=0}^\infty u_j\big|_{x_n=0}.$$

In Section 1.5 we will show that this definition can be used on $\mathbb{T} \times \mathbb{R}^n$ as well and Section 3.4 will show that the sum above indeed converges in a reasonable sense in $\mathbb{T} \times \mathbb{R}^{n-1}$ and coincides with the value of $u|_{\mathbb{T} \times \mathbb{R}^{n-1}}$, *i.e.*, $T_n u = u|_{\mathbb{T} \times \mathbb{R}^{n-1}}$. For the convergence result some assumptions on the regularity parameter s have to be made, which depend on the integrability p and anisotropy $\vec{a}$ and coincide with the ones needed in $\mathbb{R}^n$. To conclude surjectivity of the trace operator, we construct an extension operator in Section 3.3, *i.e.*, we are concerned with the following question: Given an element v in a Triebel-Lizorkin space on $\mathbb{T} \times \mathbb{R}^{n-1}$ is there an element $u = Ev$ on $\mathbb{T} \times \mathbb{R}^n$ such that $u|_{x_n=0} = v$? This is possible for all values $s \in \mathbb{R}$. The combination of these results comprehensively solves the trace problem of time-periodic parabolic problems in the half space.

In the last chapter of this thesis we apply the presented ideas of Kyed and Galdi and the theory of Chapter 3 to the time-periodic equations of magnetohydrodynamics in $\mathbb{T} \times \Omega = \Omega_\mathbb{T}$ given by

(MHDE)
$$\begin{cases} \partial_t u - \nu \Delta u + \nabla \mathfrak{p} + \dfrac{1}{2}\nabla |H|^2 + (u \cdot \nabla)u = F + (H \cdot \nabla)H & \text{in } \Omega_\mathbb{T}, \\[2mm] \partial_t H - \mu \Delta H = \nabla \times [u \times H] & \text{in } \Omega_\mathbb{T}, \\[2mm] \operatorname{div} u = \operatorname{div} H = 0 & \text{in } \Omega_\mathbb{T}, \\[2mm] u = 0, \quad H \cdot n = B_1, \quad \operatorname{curl} H \times n = 0 & \text{on } \partial\Omega_\mathbb{T}. \end{cases}$$

The factor S was omitted because it can easily be absorbed into the magnetic field H by the transformation $(u, H) \mapsto (u, S^{\frac{1}{2}}H)$. To the knowledge of the author previous considerations of this problem have been restricted to L_2 and $B_1 = 0$. We will give a few examples. Notte, Rojas and Rojas [68] considered the problem with Dirichlet boundary conditions for u and H with an additional term on the right-hand side in the equation for H related to the motion of heavy ions. Ibrahim, Lemarié Rieusset and Masmoudi [48] considered a slightly more complicated system with additional equations for E and applied the ideas of Kyed to construct time-periodic solutions. Additionally, the equations were also considered as an optimal control problem, see Gunzburger and Trenchea [43].

We proceed in Chapter 4 in the following way. As a first step we construct an extension H_0 to the boundary value B_1, *i.e.*, a function H_0 such that $H_0 \cdot n = B_1$, $\operatorname{curl} H_0 \times n = 0$ on $\partial\Omega$ and $\operatorname{div} H_0 = 0$ in Ω. With this extension we transform the system into one with homogeneous boundary data. As a first step we consider the linearisation given by

$$\begin{cases} \partial_t u - \nu \Delta u - (H_0 \cdot \nabla)H - (H \cdot \nabla)H_0 + \nabla \mathfrak{p} = F & \text{in } \Omega_\mathbb{T}, \\[2mm] \partial_t H - \mu \Delta H - \nabla \times [u \times H_0] = G & \text{in } \Omega_\mathbb{T}, \\[2mm] \operatorname{div} u = \operatorname{div} H = 0 & \text{in } \Omega_\mathbb{T}, \\[2mm] u = 0, \quad H \cdot n = 0, \quad \operatorname{curl} H \times n = 0 & \text{on } \partial\Omega_\mathbb{T}. \end{cases}$$

By perturbation theory we identify functions spaces such that for every (F, G) we obtain a solution $(u, H, \mathfrak{p})$ that satisfies a corresponding a priori estimate, where the constant $c(\kappa)$ is uniform with respect to $H_0 \in \mathrm{W}_\infty^1(\Omega)$ such that $\|H_0\|_{\mathrm{W}_\infty^1(\Omega)} \leq \kappa$. This type of behaviour of the constant is to be expected, see for example Galdi and Kyed [38, Lemma 2.4]. Using Banach's fixed-point theorem together with the stated estimate, we show existence of time-periodic solutions to (MHDE) without general smallness assumptions on B_1. Note that this includes all constant magnetic fields H_0, which is analogue to the results for the Oseen equations from [38].

Acknowledgement

I would like to express my gratitude to all people that supported me in creating this thesis.

First and foremost my advisor Professor Dr. Reinhard Farwig, who did not only give me the opportunity to be his PhD student, but also had a major influence on my undergraduate studies, which motivated me to pursue this thesis. During my time as his student, he gave me valuable support, enabled me to visit Korea and Japan and encouraged me to pursue mathematical topics that fascinated me.

Special thanks is due to Professor Dr. Mads Kyed, who piqued my interest in time-periodic problems and always had time for mathematical discussions during his time in Darmstadt. Additionally, I am very grateful that he takes the time to referee this thesis.

I owe a large thanks to my (former) colleagues of the working group Analysis, who made my time as a PhD student enjoyable, created a pleasant work environment and always found the time to discuss mathematical problems. In particular, I would like to thank Dr. Björn Augner, Sebastian Bechtel, Dr. Aday Celik, Dr. Thomas Eiter, and Andreas Schmidt for their valuable proofreading. A special thanks is due to Dr. David Wegmann, my former office mate, who helped me get started as a PhD student, and it was a pleasure to share an office with.

Last but not least, a special thanks goes to my parents for supporting me in every aspect imaginable.

Zusammenfassung in deutscher Sprache

Wir betrachten die Bewegung einer viskosen, inkompressiblen Flüssigkeit, die neutral geladen und leitfähig ist. Die Bewegung einer solchen Flüssigkeit wird durch eine Koppelung der Navier-Stokes- und Maxwell-Gleichungen beschrieben. Diese Koppelung wird oftmals magnetohydrodynamische (MHD) Gleichungen genannt und ist gegeben durch

(MHDG)
$$\begin{cases} \partial_t u - \nu \Delta u + \nabla \mathrm{p} + \frac{1}{2} \nabla |H|^2 + (u \cdot \nabla) u = F + S \cdot (H \cdot \nabla) H & \text{in } \Omega_{\mathbb{T}}, \\ \partial_t H - \mu \Delta H = \nabla \times [u \times H] & \text{in } \Omega_{\mathbb{T}}, \\ \operatorname{div} u = \operatorname{div} H = 0 & \text{in } \Omega_{\mathbb{T}}, \\ u = 0, \quad H \cdot n = B_1, \quad \operatorname{rot} H \times n = 0 & \text{auf } \partial\Omega_{\mathbb{T}}. \end{cases}$$

Dabei ist $\mathbb{T} = \mathbb{R}/\mathcal{T}\mathbb{Z}$, Ω ein beschränktes Gebiet, $\Omega_{\mathbb{T}} = \mathbb{T} \times \Omega$, F eine anliegende zeit-periodische Kraft und B_1 ein zeitunabhängiges Magnetfeld, das vom Medium des Randes erzeugt wird. Die Unbekannten hierbei sind die Geschwindigkeit der Flüssigkeit u, das Magnetfeld H und der Druck p innerhalb der Flüssigkeit.

Im ersten Teil der Arbeit erweitern wir die existierende Theorie anisotroper Triebel-Lizorkin-Räume auf zeitperiodische Funktionen. Besonders wichtig dabei ist die Theorie von Spuroperatoren von $\mathbb{T} \times \mathbb{R}^n$ nach $\mathbb{T} \times \mathbb{R}^{n-1}$ und von Fortsetzungsoperatoren von $\mathbb{T} \times \mathbb{R}^{n-1}$ nach $\mathbb{T} \times \mathbb{R}^n$. Hierbei bestimmen wir den kanonischen Spurraum anisotroper Triebel-Lizorkin-Räume, d.h. den Raum, in dem für jedes Element die Spur liegt und wir für jedes Element des Spurraumes eine Funktion finden, so dass die Spur auch angenommen wird. Diese neuen Erkenntnisse sind unerlässlich für eine umfassende Theorie von zeit-periodischen Randwertproblemen.

Der zweite Teil der Arbeit verwendet diese Resultate, um Existenz von periodischen Lösungen zu den MHD-Gleichungen in $L_q(\mathbb{T}, L_p(\Omega))$ zu zeigen. Hierbei betrachten wir zuerst eine Fortsetzung der Randdaten H_0, d.h. eine Funktion H_0 mit $H_0 \cdot n = B_1$, $\operatorname{rot} H_0 \times n = 0$ auf $\partial\Omega$ und $\operatorname{div} H_0 = 0$ in Ω und transformieren die Gleichung zu einem Problem mit homogenen Randdaten. Die Linearisierung dieses Problems ist dann gegeben durch

$$\begin{cases} \partial_t u - \Delta u - (H_0 \cdot \nabla) H - (H \cdot \nabla) H_0 + \nabla \mathfrak{p} = F & \text{in } \Omega_{\mathbb{T}}, \\ \partial_t H - \Delta H - \nabla \times [u \times H_0] = G & \text{in } \Omega_{\mathbb{T}}, \\ \operatorname{div} u = \operatorname{div} H = 0 & \text{in } \Omega_{\mathbb{T}}, \\ u = 0, \quad H \cdot n = 0, \quad \operatorname{rot} H \times n = 0 & \text{auf } \partial\Omega_{\mathbb{T}}. \end{cases}$$

A set $\Omega \subset \mathbb{R}^n$ is called a domain if it is open and connected and bounded if it is contained in $B_R(0)$ for some $R > 0$. We call a domain of class $C^{m,\alpha}$ for $m \in \mathbb{N}_0$ and $\alpha \in [0,1]$, if it has a locally finite open covering of sets $\mathcal{O}_j$ such that on each $\mathcal{O}_j$ the boundary is given as the graph of a function with $C^{m,\alpha}$-regularity. We also call these $C^{m,\alpha}$-domains or domains with a $C^{m,\alpha}$-boundary. Derivatives of sufficiently regular functions $f : \Omega \to \mathbb{C}$ will be labelled by $\partial_i f = \partial_{x_i} f$ for $i = 1, 2, \ldots, n$. ∇f denotes the gradient of f and the Laplace operator is given by $\Delta f := \sum_{i=1}^n \partial_i^2 f$. If the function has values in $\mathbb{C}^n$ we apply derivatives to each component and denote by div the divergence operator $\operatorname{div} f := \partial_j f_j$ and by curl f or rot f the rotation given by

$$\nabla \times f = \operatorname{curl} f = \begin{pmatrix} \partial_2 f_3 - \partial_3 f_2 \\ \partial_3 f_1 - \partial_1 f_3 \\ \partial_1 f_2 - \partial_2 f_1 \end{pmatrix}.$$

All of the previous operators only act on the spatial variable $x \in \Omega$. For functions $g : (0, T) \times \Omega \to \mathbb{C}^n$ we set ∂_t as the derivative with respect to the first variable, which we are going to call time from here on, and $D^\alpha g := \partial_t^{\alpha_0} \partial_1^{\alpha_1} \cdots \partial_n^{\alpha_n}$ with $\alpha = (\alpha_0, \alpha_1, \ldots, \alpha_n) \in \mathbb{N}_0^{n+1}$.

By $\lesssim$ we denote an estimate where some constant occurs that depends on the parameters but it is not important to be determined specifically. If one parameter is explicitly stated after $\lesssim$, then the constant is independent of this parameter, *i.e.*, $A \lesssim rB$ implies that the estimate $A \leq crB$ holds with a constant c independent of r.

For a vector space E with dual space E', we use $\langle u, \varphi \rangle$ to denote the dual pairing of $u \in E'$ and $\varphi \in E$. If E, F are topological vector spaces, we write $E \hookrightarrow F$ for a continuous embedding of E into F. $A \oplus B = E$ denotes the direct sum of closed subspaces $A, B \subset E$, *i.e.*, $A \cap B = \{0\}$, $A + B = E$.

We are going to use δ_{x_0} for the delta distribution on $\mathbb{R}^n$, *i.e.*, $\langle \delta_{x_0}, \varphi \rangle = \varphi(x_0)$ and δ_j for the function

$$\delta_j : \mathbb{Z} \to \mathbb{R}, \quad \delta_j(k) = \begin{cases} 1 & \text{for } k = j, \\ 0 & \text{else,} \end{cases}$$

since there is no confusion possible.

1.2 Fourier Transform on Locally Compact Abelian Groups

Since we are interested in time-periodic solutions we follow the ideas of Kyed [60] and introduce for a fixed time-period $\mathcal{T} > 0$ the torus $\mathbb{T} := \mathbb{R}/\mathcal{T}\mathbb{Z}$ and the locally compact abelian group $G_n := \mathbb{T} \times \mathbb{R}^n$. In the following we will state results from harmonic and Fourier analysis on groups, for reference on the stated results see Eiter and Kyed [30], Rudin [71], or Folland [34]. Let $[\,\cdot\,] : \mathbb{R} \to \mathbb{R}/\mathcal{T}\mathbb{Z}$ be the quotient map so that the mapping

$$\pi : \mathbb{R} \times \mathbb{R}^n \to G_n, \quad \pi(t, x) := ([t], x)$$

induces a differentiable structure and a topology on G_n. Hence, we can define

(1.1) $$C^\infty(G_n) := \left\{ f : G_n \to \mathbb{C} \mid \exists F \in C^\infty(\mathbb{R} \times \mathbb{R}^n) \text{ such that } F = f \circ \pi \right\},$$

as the space of smooth functions on G_n. Therefore, the derivatives of a function $f : G_n \to \mathbb{C}$ are defined by the derivatives of the corresponding function $f \circ \pi$, *i.e.*,

$$\partial_t^\beta \partial_x^\alpha f := \partial_t^\beta \partial_x^\alpha (f \circ \pi)\Big|_{[0,\mathcal{T}) \times \mathbb{R}^n} \qquad \text{for all } (\alpha, \beta) \in \mathbb{N}_0^n \times \mathbb{N}_0.$$

Similar to $\mathbb{R}^n$ we define for any $(\alpha, \beta, \gamma) \in \mathbb{N}_0^n \times \mathbb{N}_0 \times \mathbb{N}_0^n$ a semi-norm by

$$\rho_{\alpha,\beta,\gamma}(f) := \sup_{(t,x) \in G_n} \left| x^\gamma \partial_t^\beta \partial_x^\alpha f(t,x) \right|.$$

This creates a set of semi-norms defining the corresponding Schwartz-(Bruhat) space on G_n by

$$\mathscr{S}(G_n) := \left\{ f \in C^\infty(G_n) \mid \rho_{\alpha,\beta,\gamma}(f) < \infty \text{ for all } (\alpha,\beta,\gamma) \in \mathbb{N}_0^n \times \mathbb{N}_0 \times \mathbb{N}_0^n \right\}.$$

The countable family of semi-norm $\rho_{\alpha,\beta,\gamma}$ induces a topology on the topological vector space $\mathscr{S}(G_n)$. From classical results with respect to uniform convergence it follows that $\mathscr{S}(G_n)$ is a Fréchet space. The dual space $\mathscr{S}'(G_n)$ will be equipped with the weak-* topology, making it a locally convex sequentially complete topological vector space and its elements are called tempered distributions on G_n. By omitting the parameter β one arrives at the classical space $\mathscr{S}(\mathbb{R}^n)$ and its dual $\mathscr{S}'(\mathbb{R}^n)$, both spaces inherit the same properties as $\mathscr{S}(G_n)$ or $\mathscr{S}'(G_n)$, respectively.

Before we can define the Fourier transform, we need to state a few more properties of locally compact abelian groups. We start with the concept of the dual group of a locally compact abelian group G. It is the set of all continuous group homomorphisms from G onto the circle group, *i.e.*, the set of all complex numbers $z \in \mathbb{C}$ such that $\|z\| = 1$. We will denote the dual group by $\widehat{G}$ and call its elements characters. The dual group of $\mathbb{R}$ can be identified with $\mathbb{R}$ by $\langle \xi, x \rangle = e^{ix \cdot \xi}$. For the dual of $\mathbb{T}$ there are several options, *e.g.* $\mathbb{Z}$ by the pairing $\langle k, t \rangle = e^{i\frac{2\pi}{T} k \cdot t}$, or $\frac{2\pi}{T}\mathbb{Z}$ by the pairing $\langle k, t \rangle = e^{ik \cdot t}$. We fix $\widehat{G}_n := \mathbb{Z} \times \mathbb{R}^n$ throughout this thesis. The question of representing the bidual group is solved by the Pontryagin Duality Theorem, since it states that any locally compact abelian group naturally identifies with its bidual group. Similar to the definition of G_n we introduce the space of smooth functions on $\widehat{G}_n$ by

$$C^\infty(\widehat{G}_n) := \left\{ f : \widehat{G}_n \to \mathbb{C} \mid \forall k \in \mathbb{Z} \text{ it holds } f(k, \cdot) \in C^\infty(\mathbb{R}^n) \right\}.$$

For every $(\alpha, \beta, \gamma) \in \mathbb{N}_0^n \times \mathbb{N}_0^n \times \mathbb{N}_0$ a semi-norm is defined by

$$\hat{\rho}_{\alpha,\beta,\gamma}(f) := \sup_{(k,\xi) \in \widehat{G}_n} \left| \xi^\alpha \partial_\xi^\beta k^\gamma f(k,\xi) \right|$$

and, hence, the corresponding Schwartz-(Bruhat) space on $\widehat{G}_n$ via

$$\mathscr{S}(\widehat{G}_n) := \left\{ u \in C^\infty(\widehat{G}_n) \mid \forall (\alpha,\beta,\gamma) \in \mathbb{N}_0^n \times \mathbb{N}_0^n \times \mathbb{N}_0 : \hat{\rho}_{\alpha,\beta,\gamma}(u) < \infty \right\}.$$

Repeating the arguments from above we conclude that $\mathscr{S}(\widehat{G}_n)$ is a Fréchet space and its dual, denoted by $\mathscr{S}'(\widehat{G}_n)$, is again a locally convex sequentially complete topological vector space.

To be able to integrate functions defined on G_n, we introduce the concept of Haar measures. To every locally compact abelian group there exists a non-zero translation-invariant measure μ. It is unique up to a multiplicative constant. On G_n we normalize it such that

$$\int_{G_n} u(y) \, d\nu(y) = \frac{1}{T} \int_0^T \int_{\mathbb{R}^n} u \circ \pi(t,x) \, dx \, dt.$$

This can be done since the Lebesgue measure is a translation invariant measure and G_n and $[0, \mathcal{T}) \times \mathbb{R}^n$ generate equivalent Borel σ-algebra. By $|E|$ we denote the measure of E. We are going to omit π in the integrals for functions defined on G_n, since no confusion is possible. On $\widehat{G}_n$ the Haar measure will be normalized such that it coincides with the product of the counting measure on $\mathbb{Z}$ and the Lebesgue measure on $\mathbb{R}^n$.

The Fourier transform for a function $f : G_n \to \mathbb{C}$ is defined by

$$\mathscr{F}_{G_n}[f](k, \xi) := \widehat{f}(k, \xi) := \frac{1}{\mathcal{T}\sqrt{2\pi}^n} \int\limits_0^{\mathcal{T}} \int\limits_{\mathbb{R}^n} f(t, x) \, e^{-ix \cdot \xi - i\frac{2\pi}{\mathcal{T}} kt} \, dxdt,$$

the inverse Fourier transform, now given for a function $g : \widehat{G}_n \to \mathbb{C}$, is defined via

$$\mathscr{F}_{G_n}^{-1}[g](t, x) := g^{\vee}(t, x) := \frac{1}{\sqrt{2\pi}^n} \sum\limits_{k \in \mathbb{Z}} \int\limits_{\mathbb{R}^n} g(k, \xi) \, e^{ix \cdot \xi + i\frac{2\pi}{\mathcal{T}} kt} \, d\xi.$$

If we want to apply the Fourier transform to a function $f : \widehat{G}_n \to \mathbb{C}$ it can be defined with the general theory from above, but it is easier by the identity $\mathscr{F}_{\widehat{G}_n}[f](\cdot) := \mathscr{F}_{G_n}^{-1}[f](-\cdot)$ and similar for $\mathscr{F}_{\widehat{G}_n}^{-1}$. For a function f only dependent on space we derive $\mathscr{F}_{G_n} f = \mathscr{F}_{\mathbb{R}^n} f$ as expected and we therefore define $\mathscr{F}_{\mathbb{T}} \otimes \mathscr{F}_{\mathbb{R}^n} = \mathscr{F}_{G_n}$ and $\mathscr{F}_{\mathbb{T}}^{-1} \otimes \mathscr{F}_{\mathbb{R}^n}^{-1} = \mathscr{F}_{G_n}^{-1}$. To shorten the notation we fix $\mathscr{F} := \mathscr{F}_{G_n}$ and $\mathscr{F}^{-1} := \mathscr{F}_{G_n}^{-1}$. Therefore, if not stated otherwise, we will work with the Fourier transform on G_n. The chosen normalization implies that $\mathscr{F} : \mathscr{S}(G_n) \to \mathscr{S}(\widehat{G}_n)$ is a homeomorphism with inverse given by $\mathscr{F}^{-1}$, and by duality this extends in the usual way to $\mathscr{F} : \mathscr{S}'(G_n) \to \mathscr{S}'(\widehat{G}_n)$, i.e.,

$$\langle \mathscr{F}_{G_n} u, \varphi \rangle := \langle u, \mathscr{F}_{\widehat{G}_n} \varphi \rangle$$

for $u \in \mathscr{S}'(G_n)$ and $\varphi \in \mathscr{S}(\widehat{G}_n)$. The Pontryagin Duality Theorem is fundamental for the previous definition to be well-defined. Similar to properties of the classical Fourier transform we have the identities

$$\mathscr{F}[\partial_t^m \partial_x^\alpha u] = i^{m+|\alpha|} \left(\frac{2\pi}{\mathcal{T}} k\right)^m \xi^\alpha \mathscr{F}[u],$$

$$\partial_t^m \partial_x^\alpha \mathscr{F}^{-1}[v] = i^{m+|\alpha|} \mathscr{F}^{-1}\left[\left(\frac{2\pi}{\mathcal{T}} k\right)^m \xi^\alpha v\right]$$

for all $u \in \mathscr{S}'(G_n)$ and $v \in \mathscr{S}'(\widehat{G}_n)$. We define the convolution of $u \in \mathscr{S}'(G_n)$ and $\varphi \in \mathscr{S}(G_n)$ similar to $\mathbb{R}^n$ by

(1.2) $$[u * \varphi](t, x) := \langle u, \varphi[(t, x) - (\cdot, \cdot)] \rangle,$$

which just extends the usual definition of convolutions on locally compact groups, see Grafakos [41, Section 1.2] for details. Furthermore the well-known identities between convolution and Fourier transform also hold here, e.g. $\mathscr{F}^{-1}[u \cdot \varphi] = c_n \mathscr{F}^{-1} u * \mathscr{F}^{-1} \varphi$ for some constant $c_n > 0$. In exactly the same way we can define a convolution on $\widehat{G}_n$.

1.3 Function Spaces

In this section we will introduce most of the used function spaces and state some properties. Since these are quite standard they can be found for example in Triebel [80] and the references within. The definitions are valid for scalar and vector-valued functions. In the case of functions with values in a finite dimensional vector space any of the following norms is applied componentwise and afterwards any finite dimensional norm is applied to the norm of the components. We will make no difference in the notation between the two, since there is no confusion possible.

For topological spaces X, Y we denote the set of continuous functions $f : X \to Y$ by $\mathrm{C}(X, Y)$. If $Y = \mathbb{C}$ with the standard topology it is customary to set $\mathrm{C}(X) := \mathrm{C}(X, \mathbb{C})$ to be the vector space of bounded and continuous functions and equip it with the norm

$$\|f\|_\infty := \sup_{x \in X} |f(x)|,$$

making it a Banach space. To be able to talk about derivatives and Hölder continuity let Ω be a domain of $\mathbb{R}^n$ and X be either Ω or $\Omega_\mathbb{T} := \mathbb{T} \times \Omega$. From here on we consider functions $f : X \to \mathbb{C}$ and start by defining the spaces of m-times continuously differentiable functions by

$$\mathrm{C}^m(X) := \{ f \mid D^\alpha f \in \mathrm{C}(X) \text{ for all } |\alpha| \le m \},$$

where D^α stands for the classical derivative with $m \in \mathbb{N}_0 \cup \{\infty\}$ and $|\alpha| < \infty$. Note that the dimension of α varies depending on whether X is a subset of $\mathbb{R}^n$ or of G_n. By endowing $\mathrm{C}^m(X)$ with the norm

$$\|f\|_{\mathrm{C}^m(X)} := \sum_{|\alpha| \le m} \|D^\alpha f\|_\infty$$

it becomes a Banach space for $m < \infty$. The class of test functions on X is defined by

$$(1.3) \qquad \mathcal{D}(X) = \mathrm{C}_0^\infty(X) := \{ f \in \mathrm{C}^\infty(X) \mid \operatorname{supp} f \subset X \text{ and compact} \},$$

and its dual is given by $\mathcal{D}'(X)$. Since $\mathbb{T}$ is compact, the demand regarding the support only applies to the spatial variable. In a similar fashion we set

$$\mathcal{D}(\overline{X}) = \mathrm{C}_0^\infty(\overline{X}) := \{ f|_X \mid f \in \mathcal{D}(G_n) \};$$

the above definition works for time-periodic and time-independent functions since every function in $\mathcal{D}(\mathbb{R}^n)$ can be extended to an element of $\mathcal{D}(G_n)$ by a constant extension. For a finer scale of regularity we define the Hölder seminorm for $\alpha \in (0, 1]$ by

$$\|f\|_{\widehat{\mathrm{C}}^\alpha(X)} := \sup_{\substack{x,y \in X \\ x \ne y}} \frac{|f(x) - f(y)|}{|x - y|^\alpha}.$$

With this we introduce the Hölder-(Zygmund)-spaces $\mathrm{C}^{m,\alpha}(X)$ as the set of functions $f \in \mathrm{C}^m(X)$ such that the norm

$$\|f\|_{\mathrm{C}^{m,\alpha}(X)} := \|f\|_{\mathrm{C}^m(X)} + \sum_{|\gamma|=m} \|D^\gamma f\|_{\widehat{\mathrm{C}}^\alpha(X)} < \infty.$$

We allow for $\alpha = 1$ to include the spaces of Lipschitz continuous functions, but since we do not consider double differences, as it is done in Zygmund spaces, we put brackets in the name of the spaces above and will omit it from now on.

We continue with the consideration of function spaces on a σ-finite measure space X with measure μ to define Lebesgue and Sobolev spaces. We tacitly identify elements of these spaces, which are equivalence classes, with a representative and hence call elements functions as well. Therefore, the space $\mathrm{L}_p(X)$ is the set of measurable functions $f : X \to \mathbb{C}$ such that

$$\|f\|_{\mathrm{L}_p(X)} := \left(\int_X |f(x)|^p \, \mathrm{d}\mu(x) \right)^{\frac{1}{p}}$$

is finite in the case of $p \in [1, \infty)$. Often we will omit μ and simply write $\mathrm{d}x$ when the underlying measure is clear. For $p = \infty$ the following has to be finite

$$\|f\|_{\mathrm{L}_\infty(X)} := \operatorname*{ess\,sup}_{x \in X} |f(x)|.$$

If the underlying measure space is clear, we shorten to $\| \cdot \|_p$ and $\| \cdot \|_\infty$. Note that in general there is no confusion whether the $\| \cdot \|_\infty$ denotes the essential supremum or just the supremum. For $1 \leq p < \infty$ the weak $\mathrm{L}_{p;\infty}(X)$ is the set of measurable functions such that

$$\|f\|_{\mathrm{L}_{p;\infty}(X)} := \inf \left\{ C > 0 \mid \mu(\{x \in X \mid |f(x)| > \alpha\}) \leq \frac{C^p}{\alpha^p} \text{ for all } \alpha > 0 \right\}$$

is finite. From this definition it is clear that weak $\mathrm{L}_\infty(X)$ is defined as the classical $\mathrm{L}_\infty(X)$-space. If $X = \mathbb{Z}$ or some subset of it, we equip X with the counting measure and denote it by $\ell_p(X)$. So these spaces consist of sequences, whose p-th power is summable or are bounded in the case of $p = \infty$. An important property of these spaces is the monotonicity with respect to the power of summation, it holds $\ell_p \hookrightarrow \ell_q$ for $1 \leq p \leq q \leq \infty$.

In the case of strongly measurable functions with values in a Banach space $\mathcal{B}$, the norms are just applied to the real valued function $\|f(\cdot)\|_\mathcal{B}$, *i.e.*, $\|f\|_{\mathrm{L}_p(X;\mathcal{B})} := \|\|f\|_\mathcal{B}\|_{\mathrm{L}_p(X)}$, hence $\mathrm{L}_p(X, \mathcal{B})$ is defined in the same way as before. Regarding measurability we refer to Yosida [88, Chapter V], but note that if $\mathcal{B}$ is any of the ℓ_p-spaces the question reduces to measurability of each component of the sequence.

We extend the definition of L_p to an integrability parameter $\vec{p} \in [1, \infty]^n$ for a space $X = \prod_{i=1}^n X_i$, where each X_i is a σ-finite measure space. These spaces were introduced by Benedek and Panzone [13] and if not stated otherwise proofs of the following statements can be found in the cited paper. The space $\mathrm{L}_{\vec{p}}(X)$ with $\vec{p} \in [1, \infty]^n$ is the set of measurable functions f such that the norm

$$\|f\|_{\mathrm{L}_{\vec{p}}(X)} := \left(\int_{X_1} \cdots \left(\int_{X_{n-1}} \left(\int_{X_n} |f(x_1, \ldots, x_{n-1}, x_n)|^{p_n} \mathrm{d}x_n \right)^{\frac{p_{n-1}}{p_n}} \mathrm{d}x_{n-1} \right)^{\frac{p_{n-2}}{p_{n-1}}} \cdots \mathrm{d}x_1 \right)^{\frac{1}{p_1}}$$

is finite, here obvious modifications in the case of $p_i = \infty$ need to be made. For strongly measurable Banach space valued functions we define the norm in the same way, as we did for L_p. Before we state some properties, we introduce some notation. By $\frac{1}{\vec{p}}$ we denote the vector $\left(\frac{1}{p_1}, \frac{1}{p_2}, \ldots, \frac{1}{p_n} \right)$, where we define $\frac{1}{\infty} = 0$. Furthermore, an equality or inequality of vectors is to be understood as an equality or inequality in each component, *i.e.*, $\frac{1}{\vec{p}} + \frac{1}{\vec{q}} = 1$ means $\frac{1}{p_i} + \frac{1}{q_i} = 1$ for every component $1 \leq i \leq n$ of every $\vec{p}, \vec{q} \in [1, \infty]^n$. Hence, we have Hölder's inequality in the form

$$(1.4) \qquad \|fg\|_{\mathrm{L}_{\vec{r}}(X)} \leq \|f\|_{\mathrm{L}_{\vec{p}}(X)} \|g\|_{\mathrm{L}_{\vec{q}}(X)} \quad \text{for} \quad \frac{1}{\vec{p}} + \frac{1}{\vec{q}} = \frac{1}{\vec{r}},$$

and in the case where convolution is possible we obtain Young's inequality via

$$(1.5) \qquad \|f * g\|_{L_{\vec{r}}(X)} \leq \|f\|_{L_{\vec{p}}(X)} \|g\|_{L_{\vec{q}}(X)} \quad \text{for} \quad \frac{1}{p} + \frac{1}{q} = 1 + \frac{1}{r}.$$

We briefly remark that this includes $X_i = \mathbb{T}$ for some i in both cases, see (1.2) for the definition of convolution. The theory of Fourier multiplier is well-studied for $L_p(\mathbb{R}^n)$, see Grafakos [41, Section 6.2] for example. We will state the following theorem due to Lizorkin [65, Corollary 1].

Theorem 1.3.1. *Let $m : \mathbb{R}^n \to \mathbb{C}$ be a bounded function with continuous derivatives that satisfy*

$$|\xi^\alpha D^\alpha m(\xi)| \leq M$$

for some constant $M > 0$, all $\xi \in \mathbb{R}^n_*$, and all $\alpha \in \{0,1\}^n$. Then m is an $L_{\vec{p}}(\mathbb{R}^n)$-multiplier and there exists a constant $C > 0$ such that

$$\|\mathscr{F}^{-1}_{\mathbb{R}^n} m \mathscr{F}_{\mathbb{R}^n} f\|_{L_{\vec{p}}(\mathbb{R}^n)} \leq CM \|f\|_{L_{\vec{p}}(\mathbb{R}^n)}$$

for all $1 < \vec{p} < \infty$ and all $f \in \mathscr{S}(\mathbb{R}^n)$.

We collect some results regarding completeness and duality in the following lemma.

Lemma 1.3.2. *The spaces $L_{\vec{q}}(X)$ are Banach spaces for every $1 \leq \vec{q} \leq \infty$. Let $1 \leq \vec{p} < \infty$ and $\frac{1}{\vec{p}} + \frac{1}{\vec{p}'} = 1$. Every linear continuous functional T on $L_{\vec{p}}(X)$ can be represented by a uniquely determined $h \in L_{\vec{p}'}(X)$ such that*

$$T(f) = \int_X f(x) h(x) \, dx$$

and $\|T\| = \|h\|_{L_{\vec{p}'}(X)}$.

By this the classical duality result of L_p extends to $L_{\vec{p}}(X)^* = L_{\vec{p}'}(X)$ with above stated identity for $\vec{p}$ and $\vec{p}'$. By induction the theorems of monotone and dominated convergence as well as the Lemma of Fatou are also valid for these spaces. Therefore, it is natural to view $L_{\vec{p}}$ as an extension of L_p. Hence these spaces can be considered simultaneously, if the domain has the needed structure. This implies that if Ω does not have product structure, then the parameter $\vec{p}$ of $L_{\vec{p}}$ is to be understood as a scalar, *i.e.*, $\vec{p} = p$ for some $p \in [1, \infty]$. Hence, we can consider arbitrary domains and collect some density results.

Lemma 1.3.3. *Let $\mathcal{B}$ be a Banach space, $1 \leq \vec{p} < \infty$, and X a σ-finite measure space with measure μ.*

a) *The set $\left\{ \sum_{j=1}^m \chi_{E_j} u_j \mid u_j \in \mathcal{B}, \ E_j \subset X \text{ are pairwise disjoint and } \mu(E_j) < \infty, \ m \in \mathbb{N} \right\}$ of simple functions is dense in $L_q(X, \mathcal{B})$ for $1 \leq q < \infty$.*

b) *The set of bounded measurable functions with compact support, denoted by $L^0_\infty(\mathbb{R}^n, \mathcal{B})$, is dense in $L_{\vec{p}}(\mathbb{R}^n, \mathcal{B})$.*

c) *The spaces $\mathcal{D}(G_n)$ and $\mathscr{S}(G_n)$ are dense in $L_{\vec{p}}(G_n)$.*

is finite. For a definition of these spaces by interpolation, we refer to Adams and Fournier [1, Chapter 7]. As long as the parameter $s < 1$ the definition above applies to $\mathrm{W}_p^s(\partial\Omega)$ as well, in the case of $s > 1$ one has to define these spaces using charts and an atlas. An important result regarding these spaces concerns the trace of functions. It is known that the trace $\gamma(f) := u\big|_{\partial\Omega}$ is an element of $\mathrm{W}_p^{1-\frac{1}{p}}(\partial\Omega)$ for any function $f \in \mathrm{W}_p^1(\Omega)$ and $1 < p < \infty$. The space $\mathrm{W}_{p'}^{-s}(\partial\Omega)$ denotes the dual space of $\mathrm{W}_p^s(\partial\Omega)$ for non integer values $s > 0$. We close the theory of Sobolev spaces with a density result.

Lemma 1.3.6. *For $1 \le q, \vec{p} < \infty$ the spaces $\mathcal{D}(G_n)$ and $\mathscr{S}(G_n)$ are dense in $\mathrm{W}_{q,\vec{p}}^{l,\vec{m}}(G_n)$ for all $l \in \mathbb{N}_0$ and $\vec{m} \in \mathbb{N}_0^n$.*

Proof: From the proof of Lemma 1.3.3 we know that $f \in \mathrm{L}_{q,\vec{p}}(G_n)$ convoluted with an approximate identity φ_n converges to f in $\mathrm{L}_{q,\vec{p}}(G_n)$. From $D^\alpha[f * \varphi_n] = [D^\alpha f] * \varphi_n$ as long as $D^\alpha f \in \mathrm{L}_{q,\vec{p}}(G_n)$, we obtain density of $\mathrm{C}^\infty(G_n) \cap \mathrm{W}_{q,\vec{p}}^{l,\vec{m}}(G_n)$. A multiplication with a sequence of smooth cut-off functions with increasing and exhausting supports yields the result by the dominated convergence theorem. $\square$

Remark 1.3.7. We introduced all spaces above with a periodicity in the first variable, hence all spaces are defined on $\mathbb{T} \times \Omega$. By substituting $\mathbb{T}$ for any interval $(0, T)$ one arrives at the more classical function spaces, which inhere the same properties.

1.3.1 Projected Subspaces

An idea introduced by Kyed [60] to deal with time-periodic problems in the framework of $\mathbb{T}$ is to decompose the problem into a steady-state part, which is entirely time-independent, and a purely periodic part. The advantage is that the latter part possesses better properties by having mean value zero with respect to time. We follow his idea and, for a function $f \in \mathrm{L}_{1,\mathrm{loc}}(\Omega_\mathbb{T})$, introduce the projections

$$(1.7) \qquad \mathcal{P}f := \frac{1}{T} \int_\mathbb{T} f(t, x)\, \mathrm{d}t, \quad \mathcal{P}_\perp f := f - \mathcal{P}f.$$

Since these projections continuously map $\mathcal{D}(\Omega_\mathbb{T})$ into $\mathcal{D}(\Omega_\mathbb{T})$ and $\mathscr{S}(G_n)$ into $\mathscr{S}(G_n)$, they can be extended to continuous projections on $\mathcal{D}'(\Omega_\mathbb{T})$ and $\mathscr{S}'(G_n)$ by

$$\langle \mathcal{P}u, \varphi \rangle := \langle u, \mathcal{P}\varphi \rangle$$

for $u \in \mathcal{D}'(\Omega_\mathbb{T})$ and $\varphi \in \mathcal{D}(\Omega_\mathbb{T})$ respectively $u \in \mathscr{S}'(\Omega_\mathbb{T})$ and $\varphi \in \mathscr{S}(\Omega_\mathbb{T})$. By making use of the Fourier transform on the torus we conclude the identity $\mathcal{P}\varphi = \mathscr{F}_\mathbb{T}^{-1}\delta_0 \mathscr{F}_\mathbb{T}\varphi$ and hence the definitions

$$(1.8) \qquad \mathcal{P}u := \mathscr{F}_\mathbb{T}^{-1}\delta_0 \mathscr{F}_\mathbb{T}u, \quad \mathcal{P}_\perp u := \mathscr{F}_\mathbb{T}^{-1}\big[(1 - \delta_0)\mathscr{F}_\mathbb{T}u\big]$$

for $u \in \mathcal{D}'(\Omega_\mathbb{T})$ or $u \in \mathscr{S}'(G_n)$. We denote $\mathcal{P}_\perp u$ as the purely periodic part and $\mathcal{P}u$ the stationary part. In the case of Lebesgue or Sobolev spaces it is clear that the projections are continuous and generate a direct decomposition into a purely periodic part of the function space which we will denote by the subscript $\perp$ and in a stationary part, *i.e.*, it holds $\mathrm{L}_{q,\vec{p}}(\Omega_\mathbb{T}) = \mathrm{L}_{q,\vec{p},\perp}(\Omega_\mathbb{T}) \oplus \mathrm{L}_{\vec{p}}(\Omega)$, where $\mathrm{L}_{q,\vec{p},\perp}(\Omega_\mathbb{T}) := \mathcal{P}_\perp \mathrm{L}_{q,\vec{p}}(\Omega_\mathbb{T})$, with similar notation and results for Sobolev spaces $\mathrm{W}_{q,\vec{p}}^{l,\vec{m}}(\Omega_\mathbb{T})$. The advantage of this decomposition will be seen in Section 4.2.2.

An important projection when dealing with incompressible fluid dynamics is the Helmholtz projection denoted by $\mathcal{P}_H$. In the following we collect properties the proofs of which can be found in Galdi [35, Chapter III] for example. The spaces of solenoidal test functions will be denoted by

$$C_{0,\sigma}^\infty(\Omega) := \left\{ \varphi \in C_0^\infty(\Omega)^n \mid \operatorname{div} \varphi = 0 \right\}.$$

The corresponding L_p spaces are given by

$$L_{p,\sigma}(\Omega) := \overline{C_{0,\sigma}^\infty(\Omega)}^{\|\cdot\|_{L_p}}, \quad \mathcal{G}_p(\Omega) := \left\{ \nabla \mathfrak{p} \mid \mathfrak{p} \in \widehat{W}_p^1(\Omega) \right\}.$$

For time-dependent functions we define $L_{q,p,\sigma}(\Omega_{\mathbb{T}}) := L_q(\mathbb{T}, L_{p,\sigma}(\Omega))$. Lemma 1.3.4 implies that the space coincides with the closure of $C_{0,\sigma}^\infty(\Omega_{\mathbb{T}}) := C^\infty(\mathbb{T}, C_{0,\sigma}^\infty(\Omega))$ in the $L_{q,p}(\Omega_{\mathbb{T}})$-norm for $1 \le q, p < \infty$.

Lemma 1.3.8. *Let Ω be a bounded domain of class C^2, the half-space $\mathbb{R}_+^n$ or the whole $\mathbb{R}^n$. For every $1 < p < \infty$ the Helmholtz projection $\mathcal{P}_H : L_p(\Omega)^n \to L_{p,\sigma}(\Omega)$ exists and is continuous. Furthermore $L_p(\Omega)^n = L_{p,\sigma}(\Omega) \oplus \mathcal{G}_p(\Omega)$ holds as a direct decomposition.*

The previous Lemma holds true for any domain in the case of $p = 2$ and it is possible to weaken the restriction on the regularity of the domain in the other cases. For any function $u \in L_p(\Omega)$ with $\operatorname{div} u \in L_p(\Omega)$ the trace $u \cdot n$ can be defined as an element of $W_p^{-\frac{1}{p}}(\partial\Omega)$ by

$$(1.9) \qquad \langle u \cdot n, \varphi \rangle_{\partial\Omega} := \int_\Omega u(x) \cdot \nabla\varphi(x) \, \mathrm{d}x + \int_\Omega \operatorname{div} u(x) \varphi(x) \, \mathrm{d}x$$

for all $\varphi \in W_{p'}^1(\Omega)$. This is justified by the fact every u with the above properties can be approximated by elements of $\mathcal{D}(\overline{\Omega})$ and for these the identity holds by an application of the divergence theorem. Motivated by (1.9) Galdi [35, Theorem III.2.3] showed the following identity

$$(1.10) \qquad L_{p,\sigma}(\Omega) = \left\{ u \in L_p(\Omega) \mid \operatorname{div} u = 0 \text{ in } \Omega, \ u \cdot n = 0 \text{ in } W_p^{-\frac{1}{p}}(\partial\Omega) \right\}.$$

Additionally he showed, see [35, Lemma III.2.1], that a vector field $u \in L_p(\Omega)^n$ is in $L_{p,\sigma}(\Omega)$ for $1 < p < \infty$ if and only if

$$(1.11) \qquad \int_\Omega u(x) \cdot \nabla h(x) \, \mathrm{d}x = 0$$

for all $h \in \widehat{W}_{p'}^1(\Omega)$.

1.4 Anisotropic Distance Function

In this section we are going to define an anisotropic distance function. The idea goes back to Fabes and Rivière [32], was extended by Yamazaki [85], and has become one of the standard tools to define anisotropic functions spaces, see for example Yamazaki [85,86], Johnsen and Sickel [55,56] or Georgiadis and Nielsen [39]. Because the proofs of properties of the anisotropic distance function are a bit sparse in the literature but often quite direct, we will state and prove the needed ones.

Definition 1.4.1. Let $\vec{a} \in (0, \infty)^n$ be given. For $t \in \mathbb{R}$ with $t \geq 0$, and $x \in \mathbb{R}^n$ we define $t^{\vec{a}} x := (t^{a_1} x_1, t^{a_2} x_2, \ldots, t^{a_n} x_n)$ and $t^{s\vec{a}} x := (t^s)^{\vec{a}} x$ for any $s \in \mathbb{R}$. The function $|\cdot|_{\vec{a}} : \mathbb{R}^n \to \mathbb{R}^+$ is defined by the unique $t \in \mathbb{R}_+$ such that $t^{-\vec{a}} x \in S^{n-1}$, *i.e.*,

$$(1.12) \qquad \sum_{j=1}^{n} \frac{x_j^2}{|x|_{\vec{a}}^{2a_j}} = 1$$

for any $x \neq 0$, and $|0|_{\vec{a}} := 0$. Furthermore, we define $|\vec{a}| = \sum_{j=1}^{n} a_j$ as the length of the anisotropy.

It is easy to see that in the case of $\vec{a} = 1$ we have the euclidean norm. Since the function $t \mapsto \sum_{k=1}^{n} \frac{x_k^2}{t^{2a_k}}$ is continuous, strictly decreasing and tends to 0 for $t \to \infty$ and to ∞ for $t \to 0$ the above definition is well defined. Let us now state and prove elementary properties of $|\cdot|_{\vec{a}}$; for this we need

$$(1.13) \qquad \tau := \min\{1, a_1, a_2, \ldots, a_n\}.$$

Proposition 1.4.2. *The anisotropic distance function $|\cdot|_{\vec{a}} : \mathbb{R}^n \to \mathbb{R}_+$ has for every $\vec{a} \in (0, \infty)^n$ the following properties:*

a) $|t^{\vec{a}} x|_{\vec{a}} = t|x|_{\vec{a}}$ *for every $x \in \mathbb{R}^n$,*

b) It holds $|x|_{\vec{a}} = |x|_{\vec{b}}^{\lambda}$ for $\lambda \vec{a} = \vec{b}$,

c) $|x + y|_{\vec{a}} \leq 2^{\frac{1}{\tau} - 1}(|x|_{\vec{a}} + |y|_{\vec{a}})$,

d) $\max_j\{|x_j|^{\frac{1}{a_j}}\} \leq |x|_{\vec{a}} \leq n^{\frac{1}{\tau} - 1} \sum_{j=1}^{n} |x_j|^{\frac{1}{a_j}}$,

e) $|\cdot|_{\vec{a}} \in C^{\infty}(\mathbb{R}^n \setminus \{0\})$ *and to every $\alpha \in \mathbb{N}_0^n$ and every $s \in \mathbb{R}$ there exists a constant $c(s, \alpha, \vec{a})$ such that*

$$|D^{\alpha} |x|_{\vec{a}}^s| \leq c(s, \alpha, \vec{a}) |x|_{\vec{a}}^{s - \alpha \cdot \vec{a}},$$
$$|x^{\alpha} D^{\alpha} |x|_{\vec{a}}^s| \leq c(s, \alpha, \vec{a}) |x|_{\vec{a}}^s.$$

Proof: Statements a) and b) are clear for $x = 0$. For $x \neq 0$ we have $(t^{-2})^{\vec{a}} t^{\vec{a}} x \in S^{n-1}$ if and only if $t^{-\vec{a}} x \in S^{n-1}$ and therefore a) follows. For $\lambda \vec{a} = \vec{b}$ we derive

$$\sum_{k=1}^{n} \frac{x_k^2}{\left(|x|_{\vec{b}}^{\lambda}\right)^{2a_k}} = \sum_{k=1}^{n} \frac{x_k^2}{|x|_{\vec{b}}^{2a_k\lambda}} = \sum_{k=1}^{n} \frac{x_k^2}{|x|_{\vec{b}}^{2b_k}} = 1,$$

hence $|x|_{\vec{a}} = |x|_{\vec{b}}^{\lambda}$ and thus b).

To prove c) we first consider the case of $\vec{a} \geq 1$ and want to show that

$$\sum_{k=1}^{n} \frac{(x_k + y_k)^2}{(|x|_{\vec{a}} + |y|_{\vec{a}})^{2a_k}} \leq 1,$$

since this would imply $|x+y|_{\vec{a}} \leq |x|_{\vec{a}} + |y|_{\vec{a}}$. Note that because by a) the sum is invariant under the transformation $(x, y) \to (t^{\vec{a}} x, t^{\vec{a}} y)$, we can assume that $|x|_{\vec{a}} + |y|_{\vec{a}} = 1$ and therefore $|x|_{\vec{a}}^{2a_k} \leq |x|_{\vec{a}}^2$ for all k. This yields

$$1 = \sum_{k=1}^{n} \frac{x_k^2}{|x|_{\vec{a}}^{2a_k}} \geq \sum_{k=1}^{n} \frac{x_k^2}{|x|_{\vec{a}}^2} = \frac{|x|^2}{|x|_{\vec{a}}^2}$$

and hence $|x + y| \leq |x| + |y| \leq |x|_{\vec{a}} + |y|_{\vec{a}} = 1$. Therefore,

$$\sum_{k=1}^{n} \frac{(x_k + y_k)^2}{(|x|_{\vec{a}} + |y|_{\vec{a}})^{2a_k}} = |x + y|^2 \leq 1$$

and hence $|x + y|_{\vec{a}} \leq |x|_{\vec{a}} + |y|_{\vec{a}}$ for all $\vec{a} \geq 1$. For a given $\vec{a} \in (0, \infty)^n$ we find $\vec{b} \in [1, \infty)^n$ such that $\tau \vec{b} = \vec{a}$. Thus we obtain from b) with $\lambda = \frac{1}{\tau}$

(1.14) $$|x + y|_{\vec{a}}^{\tau} = |x + y|_{\vec{b}}^{\tau} \leq |x|_{\vec{b}}^{\tau} + |y|_{\vec{b}}^{\tau} = |x|_{\vec{a}}^{\tau} + |y|_{\vec{a}}^{\tau} \leq 2^{1-\tau}(|x|_{\vec{a}} + |y|_{\vec{a}})^{\tau}$$

and hence c).

The estimate to the left in d) holds since at each $|x_j|^{\frac{1}{a_j}}$ the sum from (1.12) is at least 1. The estimate to the right follows from c) for $\vec{a} \geq 1$, and in the case of $\vec{a} \in (0, \infty)^n$ we apply the ideas from equation (1.14) to obtain the stated constant.

By the implicit function theorem the function $|\cdot|_{\vec{a}}$ has the stated regularity. Since $|t^{\vec{a}} x|_{\vec{a}}^s = t^s |x|_{\vec{a}}^s$ from a) and $[D^{\alpha} |\cdot|_{\vec{a}}^s](t^{\vec{a}} x) = t^{s - \alpha \cdot \vec{a}} D^{\alpha} |x|_{\vec{a}}^s$ by Lemma 5.1.2 we take $t = |x|_{\vec{a}}^{-1}$ and conclude

$$\frac{D^{\alpha} |x|_{\vec{a}}^s}{|x|_{\vec{a}}^{s - \alpha \cdot \vec{a}}} = \frac{D^{\alpha} \left| \frac{x}{|x|_{\vec{a}}} \right|_{\vec{a}}^s}{\left| \frac{x}{|x|_{\vec{a}}} \right|_{\vec{a}}^{s - \alpha \cdot \vec{a}}}.$$

On the compact set $\{x \in \mathbb{R}^n \mid |x|_{\vec{a}} = 1\}$ the denominator and numerator are bounded, because both functions are continuous by the previous argument. Furthermore the function $|\cdot|_{\vec{a}}$ is bounded from below, hence, the quotient is bounded as well. This implies the existence of a constant $c(s, \alpha, \vec{a})$ by the stated scaling. Therefore, the first estimate of e) is proven. The second estimate follows from d), because $|x^{\alpha}| \leq |x|_{\vec{a}}^{\alpha \cdot \vec{a}}$. $\qquad \square$

Remark 1.4.3. It is quite easy to see, that the function $|\cdot|_{\vec{a}}$ is in fact not regular in 0 if at least one $a_j \neq 1$. For example for $a_1 = 2$ it holds $|(x_1, 0)|_{\vec{a}} = \sqrt{|x_1|}$ and hence an irregularity at $x_1 = 0$ arises. Furthermore, in the case of one $a_j < 1$ the constant in part c) and d) is necessary and in fact optimal since $|e_j + e_j|_{\vec{a}} = 2^{\frac{1}{a_j}} \leq 2^{\frac{1}{\tau} - 1}(|e_j|_{\vec{a}} + |e_j|_{\vec{a}})$ with equality in the case of $a_j = \tau$.

To be able to use some form of anisotropic function as an operator on $\mathscr{S}'(\mathbb{R}^n)$ we define the following.

Lemma 1.4.4. *The function $\langle \cdot \rangle_{\vec{a}} : \mathbb{R}^n \to \mathbb{R}$ given by $\langle x \rangle_{\vec{a}} := |(1, x)|_{(1, \vec{a})}$ is in $C^{\infty}(\mathbb{R}^n)$ and for $s \in \mathbb{R}$ and $\alpha \in \mathbb{N}_0^n$ there exist constants $C(s, \alpha, \vec{a})$ so that*

$$|D^{\alpha} \langle x \rangle_{\vec{a}}^s| \leq C(s, \alpha, \vec{a}) \langle x \rangle_{\vec{a}}^{s - \alpha \cdot \vec{a}},$$
$$|x^{\alpha} D^{\alpha} \langle x \rangle_{\vec{a}}^s| \leq C(s, \alpha, \vec{a}) \langle x \rangle_{\vec{a}}^s,$$
$$\langle t^{\vec{a}} x \rangle_{\vec{a}} \leq t \langle x \rangle_{\vec{a}} \quad for \ t \geq 1.$$

Proof: By Proposition 1.4.2 e) we have that $\langle \cdot \rangle_{\vec{a}} \in C^{\infty}(\mathbb{R}^n)$ because $(1, x)$ never reaches the origin. Additionally, from $\langle x \rangle_{\vec{a}} = |(1, x)|_{(1, \vec{a})}$ and $|x|_{\vec{a}} \leq \langle x \rangle_{\vec{a}}$ part e) also yields the first two estimates. The third one follows from $\langle t^{\vec{a}} x \rangle = |(1, t^{\vec{a}} x)|_{(1, \vec{a})} \leq |(t, t^{\vec{a}} x)|_{(1, \vec{a})} = t \langle x \rangle$, since $t \geq 1$ and $|\cdot|_{\vec{a}}$ is monotone. $\qquad \square$

Proof: By the well-known theorem of Paley-Wiener-Schwartz, see Hörmander [46, Theorem 7.3.1], it follows immediately that the Fourier transform maps $\mathcal{H}'_{(M,K)}$ onto $\mathcal{S}'_{(M,K)}(\widehat{G}_n)$ since exactly all integrals with respect to the torus do not vanish if $k \in M$.

Now let $u \in \mathcal{S}'_{(M,K)}$. Lemma 1.5.9 implies $u = \sum_{j \in M} u_j \otimes \delta_j$ because the rest of the sum is zero by the assumption on the support of u. From the definition of u_j we obtain $\operatorname{supp} u_j \subset K$ for every $j \in M$. Hence the classical theorem implies that $f_j := \mathcal{F}_{\mathbb{R}^n}^{-1}(u_j)$ has the mentioned properties of the theorem. Since there are only finitely many u_j, we indeed derive global constants $C > 0$ and $n \in \mathbb{N}_0$. The result now follows from

$$\mathcal{F}_{G_n}^{-1} u = \sum_{j \in M} \mathcal{F}_{\mathbb{R}^n}^{-1} u_j \otimes \mathcal{F}_{\mathbb{T}}^{-1} \delta_j = \sum_{j \in M} f_j \otimes e^{i\frac{2\pi}{T} jt} = \sum_{j \in M} e^{i\frac{2\pi}{T} jt} f_j,$$

because the tensor product of two regular distributions is just the product. $\qquad\square$

The structure results answer the question how an element of $u \in \mathcal{S}'(G_n)$ can be extended to an element of $\mathcal{S}'(\mathbb{R}^{n+1})$. Lemma 1.5.10 yields the identity

$$u = \sum_{j \in \mathbb{Z}} u_j \otimes e^{i\frac{2\pi}{T} jt},$$

with $u_j = \mathcal{F}_{\mathbb{R}^n}^{-1}[(\mathcal{F}_{G_n} u)_j]$ and $(\mathcal{F}_{G_n} u)_j \in \mathcal{S}'(\mathbb{R}^n)$ defined in (1.15). Intuitively we can define

$$(1.16) \qquad\qquad u \circ \pi = \sum_{j \in \mathbb{Z}} u_j \otimes e^{i\frac{2\pi}{T} jt},$$

by just interpreting $e^{i\frac{2\pi}{T} jt}$ as an element of $\mathcal{S}'(\mathbb{R})$. Therefore, each summand is an element of $\mathcal{S}'(G_n)$ and convergence is the only point that is left to prove.

Lemma 1.5.12. *Let $u \in \mathcal{S}'(G_n)$. Then $u \circ \pi$ defined by (1.16) is well-defined and an element of $\mathcal{S}'(\mathbb{R}^{n+1})$. Furthermore we have the identities*

$$\mathcal{F}_{\mathbb{R}^{n+1}}(u \circ \pi) = \sqrt{2\pi} \sum_{j \in \mathbb{Z}} \delta_{\frac{2\pi}{T} j} \otimes (\mathcal{F} u)_j,$$

$$\mathcal{F}_{\mathbb{R}^{n+1}}^{-1}[\psi \mathcal{F}_{\mathbb{R}^{n+1}}(u \circ \pi)] = \sum_{j \in \mathbb{Z}} e^{i\frac{2\pi}{T} jt} \otimes \mathcal{F}_{\mathbb{R}^n}^{-1}[\psi\Big(\frac{2\pi}{T} j, x\Big)(\mathcal{F} u)_j],$$

for arbitrary $\psi \in \mathcal{S}(\mathbb{R} \times \mathbb{R}^n)$.

Proof: As stated we are left to prove convergence in $\mathcal{S}'(\mathbb{R}^{n+1})$. For this we note that for $\varphi \in \mathcal{S}(\mathbb{R} \times \mathbb{R}^n)$ we conclude $\langle e^{i\frac{2\pi}{T} jt}, \varphi(\cdot, x)\rangle = \sqrt{2\pi} \mathcal{F}_{\mathbb{R}}^{-1}[\varphi(\cdot, x)](\frac{2\pi}{T} j)$. Hence, for fixed $j \in \mathbb{Z}$ it holds

$$\langle e^{i\frac{2\pi}{T} jt} \otimes u_j, \varphi\rangle = \sqrt{2\pi} \Big\langle u_j, [\mathcal{F}_{\mathbb{R}}^{-1} \varphi]\Big(\frac{2\pi}{T} j, \cdot\Big)\Big\rangle = \sqrt{2\pi} \Big\langle \mathcal{F}_{\mathbb{R}^n}^{-1}[(\mathcal{F}_{G_n} u)_j], [\mathcal{F}_{\mathbb{R}}^{-1} \varphi]\Big(\frac{2\pi}{T} j, \cdot\Big)\Big\rangle$$

$$= \sqrt{2\pi} \Big\langle \mathcal{F}_{G_n} u, \delta_j(\cdot) \mathcal{F}_{\mathbb{R}^{n+1}}^{-1} \varphi\Big(\frac{2\pi}{T} j, \cdot\Big)\Big\rangle.$$

Lemma 1.5.3 implies the existence of a constant $c > 0$ and an $m \in \mathbb{N}$ independent of j such that

$$|\langle e^{i\frac{2\pi}{T} jt} \otimes u_j, \varphi\rangle| \leq c \sup_{(k,\xi) \in \widehat{G}_n} (1 + |k|)^m (1 + |\xi|)^m \Big| \delta_j(k) \hat{D}_\xi^m \mathcal{F}_{\mathbb{R}^{n+1}}^{-1} \varphi\Big(\frac{2\pi}{T} j, \xi\Big)\Big|$$

$$\leq c(1 + |j|)^{-2} \sup_{x \in \mathbb{R}^{n+1}} (1 + |x|)^{2m+2} |\hat{D}_x^m \mathcal{F}_{\mathbb{R}^{n+1}}^{-1} \varphi(x)|.$$

This estimate implies that $\sum_{|j| \leq N} u_j \otimes e^{i\frac{2\pi}{\mathcal{T}}jt}$ is a Cauchy sequence in $\mathscr{S}'(\mathbb{R}^{n+1})$ and hence convergent. Since $u \circ \pi$ is an element of $\mathscr{S}'(\mathbb{R}^{n+1})$ we can calculate the Fourier transform $\mathscr{F}_{\mathbb{R}^{n+1}} u \circ \pi$ and for $\varphi \in \mathscr{S}(\mathbb{R} \times \mathbb{R}^n)$ we have

$$\langle \mathscr{F}_{\mathbb{R}^{n+1}}(u \circ \pi), \varphi \rangle = \sum_{j \in \mathbb{Z}} \langle e^{i\frac{2\pi}{\mathcal{T}}jt} \otimes u_j, \mathscr{F}_{\mathbb{R}^{n+1}}\varphi \rangle = \sum_{j \in \mathbb{Z}} \langle e^{i\frac{2\pi}{\mathcal{T}}jt}, \langle (\mathscr{F}_{G_n}u)_j, [\mathscr{F}_{\mathbb{R}}\varphi](t,\cdot) \rangle \rangle$$

$$= \sum_{j \in \mathbb{Z}} \langle (\mathscr{F}_{G_n}u)_j, \langle e^{i\frac{2\pi}{\mathcal{T}}jt}, [\mathscr{F}_{\mathbb{R}}\varphi](t,\cdot) \rangle \rangle = \sqrt{2\pi} \sum_{j \in \mathbb{Z}} \langle \delta_{\frac{2\pi}{\mathcal{T}}j} \otimes (\mathscr{F}_{G_n}u)_j, \varphi \rangle,$$

where we used the identity from the beginning of the proof. For $\varphi \in \mathscr{S}(\mathbb{R} \times \mathbb{R}^n)$ it holds

$$\langle \mathscr{F}_{\mathbb{R}^{n+1}}^{-1} [\psi \mathscr{F}_{\mathbb{R}^{n+1}}(u \circ \pi)], \varphi \rangle = \sqrt{2\pi} \sum_{j \in \mathbb{Z}} \langle \delta_{\frac{2\pi}{\mathcal{T}}j} \otimes (\mathscr{F}u)_j, \psi \mathscr{F}_{\mathbb{R}^{n+1}}^{-1}\varphi \rangle$$

$$= \sqrt{2\pi} \sum_{j \in \mathbb{Z}} \langle (\mathscr{F}u)_j, \psi\left(\frac{2\pi}{\mathcal{T}}j, \cdot\right) [\mathscr{F}_{\mathbb{R}^{n+1}}^{-1}\varphi]\left(\frac{2\pi}{\mathcal{T}}j, \cdot\right) \rangle$$

$$= \sqrt{2\pi} \sum_{j \in \mathbb{Z}} \langle \delta_{\frac{2\pi}{\mathcal{T}}j} \otimes \left[\psi\left(\frac{2\pi}{\mathcal{T}}j, \cdot\right)(\mathscr{F}u)_j\right], \mathscr{F}_{\mathbb{R}^{n+1}}^{-1}\varphi \rangle$$

$$= \sqrt{2\pi} \sum_{j \in \mathbb{Z}} \langle \mathscr{F}_{\mathbb{R}}^{-1}\delta_{\frac{2\pi}{\mathcal{T}}j} \otimes \mathscr{F}_{\mathbb{R}^n}^{-1}\left[\psi\left(\frac{2\pi}{\mathcal{T}}j, \cdot\right)(\mathscr{F}u)_j\right], \varphi \rangle$$

$$= \sum_{j \in \mathbb{Z}} \langle e^{i\frac{2\pi}{\mathcal{T}}jt} \otimes \mathscr{F}_{\mathbb{R}^n}^{-1}\left[\psi\left(\frac{2\pi}{\mathcal{T}}j, \cdot\right)(\mathscr{F}u)_j\right], \varphi \rangle,$$

because $\sqrt{2\pi} \mathscr{F}_{\mathbb{R}}^{-1} \delta_{\frac{2\pi}{\mathcal{T}}j} = e^{i\frac{2\pi}{\mathcal{T}}jt}$. $\qquad\square$

An important consequence of the previous result is that it allows us to transfer L_∞-estimates with respect to $\mathbb{T}$. This form of extension does not provide any form of decay, thus it is clear that no other form of estimate can be transferred.

Corollary 1.5.13. *Let* $1 \leq \vec{p} \leq \infty$. *For* $u \in \mathscr{S}'(G_n)$ *and* $\psi \in \mathrm{C}_0^\infty(\mathbb{R} \times \mathbb{R}^n)$ *we have the identity*

$$(1.17) \qquad \left\| \mathscr{F}_{\mathbb{R}^{n+1}}^{-1}\left[\psi\left(\frac{\mathcal{T}}{2\pi}\cdot, \cdot\right)\mathscr{F}_{\mathbb{R}^{n+1}}(u \circ \pi)\right] \right\|_{\mathrm{L}_{\infty,\vec{p}}(\mathbb{R}^{n+1})} = \left\| \mathscr{F}^{-1}\left[\psi|_{\widehat{G}_n}\mathscr{F}u\right] \right\|_{\mathrm{L}_{\infty,\vec{p}}(G_n)}$$

Proof: Since $\mathscr{F}^{-1}[\psi|_{\widehat{G}_n}\mathscr{F}u] = \sum_{j \in \mathbb{Z}} e^{i\frac{2\pi}{\mathcal{T}}jt} \otimes \mathscr{F}_{\mathbb{R}^n}^{-1}[\psi(j,x)(\mathscr{F}u)_j]$ the identity follows directly from Lemma 1.5.12 and Theorem 1.5.11, because each summand is indeed a function and there are only finitely many entries. $\qquad\square$

1.6 Convergence Results of Distributions

In this section we discuss convergence in $\mathscr{S}'(G_n)$ of series $\sum_{j=0}^\infty u_j$, $u_j \in \mathscr{S}'(G_n)$, where the support of $\mathscr{F}_{G_n}u_j$ is contained in a compact set for each $j \in \mathbb{N}_0$. Throughout this section we take an anisotropy given by $(b, \vec{a}) \in (0, \infty)^{n+1}$. There are two types of conditions that are usually imposed, we will see applications of the different conditions in Section 3.1.1 and Section 3.2. The first is the *dyadic corona condition* and is defined by the existence of constants $A, B > 0$ such that for every $j \geq 1$ we have

$$(1.18) \qquad \mathrm{supp}\, \mathscr{F}_{G_n}u_j \subset \left\{ (k, \xi) \in \widehat{G}_n \,\middle|\, B2^{j-1} \leq |(k, \xi)|_{b,\vec{a}} \leq A2^j \right\},$$

$$\text{while } \mathrm{supp}\, \mathscr{F}_{G_n}u_0 \subset \left\{ (k, \xi) \in \widehat{G}_n \,\middle|\, |(k, \xi)|_{b,\vec{a}} \leq A \right\}.$$

The second one is the *dyadic ball condition* and is defined by the existence of a constant $A > 0$ such that for every $j \geq 0$ we have

$$(1.19) \qquad \operatorname{supp} \mathscr{F}_{G_n} u_j \subset \left\{ (k, \xi) \in \widehat{G}_n \mid |(k, \xi)|_{b, \bar{a}} \leq A 2^j \right\}.$$

To conclude convergence in $\mathscr{S}'(G_n)$ some extra estimates need to be fulfilled by u_j. Here we consider the case of pointwise estimates; a similar result in $\mathbb{R}^n$ can be found in Johnsen and Sickel [56, Lemma 3.17].

Lemma 1.6.1. *Let $(u_j)_{j \in \mathbb{N}_0}$ be a sequence of functions in $\mathscr{S}'(G_n)$ satisfying (1.18). Furthermore assume that there exists a constant $c > 0$ and an $m \in \mathbb{N}$ such that*

$$|u_j(t, x)| \leq c 2^{jm} (1 + |x|)^m \quad \text{for all } j \geq 1.$$

Then $\sum_{j=0}^{\infty} u_j$ converges in $\mathscr{S}'(G_n)$ to a distribution u.

Proof: We take $\psi \in C_0^\infty(\mathbb{R})$ so that $\operatorname{supp} \psi \subset \left\{ x \in \mathbb{R} \mid \frac{B}{4} \leq x \leq 2A \right\}$ and $\psi \equiv 1$ for $\frac{B}{2} \leq x \leq A$. Define $\psi_j : \widehat{G}_n \to \mathbb{R}$ by $\psi_j(k, \xi) := \psi(2^{-j} |(k, \xi)|_{b, \bar{a}})$ for $j \geq 0$. From the construction we conclude $u_j = \mathscr{F}^{-1} \psi_j \mathscr{F} u_j$ for $j \geq 1$. For $\varphi \in \mathscr{S}(G_n)$ we hence obtain the estimate

$$\begin{aligned}
|\langle u_j, \varphi \rangle| &= |\langle u_j, \mathscr{F} \psi_j \mathscr{F}^{-1} \varphi \rangle| = |\langle u_j, \mathscr{F}^{-1} \psi_j \mathscr{F} \varphi \rangle| \\
&\leq \left\| (1 + |x|^2)^{-m-n} u_j \right\|_{L_2(G_n)} \left\| (1 + |x|^2)^{m+n} \mathscr{F}^{-1} \psi_j \mathscr{F} \varphi \right\|_{L_2(G_n)} \\
&\lesssim 2^{jm} \left\| (1 + |x|^2)^{m+n} \mathscr{F}^{-1} \psi_j \mathscr{F} \varphi \right\|_{L_2(G_n)}.
\end{aligned}$$

Thus we are left to estimate the L_2-norm and by applying the Fourier transform we derive

$$\begin{aligned}
\left\| (1 + |x|^2)^{m+n} \mathscr{F}^{-1} \psi_j \mathscr{F} \varphi \right\|_{L_2(G_n)} &= \left\| (1 - \Delta)^{m+n} \psi_j \mathscr{F} \varphi \right\|_{L_2(\widehat{G}_n)} \\
&\lesssim \sum_{|\alpha| \leq 2n + 2m} \left\| D^\alpha (\psi_j \mathscr{F} \varphi) \right\|_{L_2(\widehat{G}_n)} \\
&\leq \sum_{\substack{|\alpha| \leq 2n + 2m \\ \gamma \leq \alpha}} \binom{\alpha}{\gamma} \left\| D^\gamma \psi_j D^{\alpha - \gamma} \mathscr{F} \varphi \right\|_{L_2(\widehat{G}_n)}.
\end{aligned}$$

To estimate further we note that $\operatorname{supp} \psi_j \subset \{ (k, \xi) \in \widehat{G}_n \mid B 2^{j-2} \leq |(k, \xi)|_{b, \bar{a}} \leq A 2^{j+1} \} =: A_j$ and

$$D^\gamma \psi_j = D^\gamma [\psi_0(2^{-jb} k, 2^{-j\bar{a}} \xi)] = 2^{-j\gamma \cdot \bar{a}} (D^\gamma \psi_0)(2^{-jb} k, 2^{-j\bar{a}} \xi).$$

This yields

$$\begin{aligned}
\left\| D^\gamma \psi_j D^{\alpha - \gamma} \mathscr{F} \varphi \right\|_{L_2(\widehat{G}_n)} &\leq 2^{-j\bar{a} \cdot \gamma} \left\| D^\gamma \psi_0 \right\|_{L_\infty(\widehat{G}_n)} \left\| D^{\alpha - \gamma} \mathscr{F} \varphi \right\|_{L_2(A_j)} \\
&\leq 2^{-j\bar{a} \cdot \gamma} \left\| D^\gamma \psi_0 \right\|_{L_\infty(\widehat{G}_n)} \left\| |(k, \xi)|_{b, \bar{a}}^l D^{\alpha - \gamma} \mathscr{F} \varphi \right\|_{L_\infty(\widehat{G}_n)} \left\| |(k, \xi)|_{b, \bar{a}}^{-l} \right\|_{L_2(A_j)}
\end{aligned}$$

for some $l \in \mathbb{N}$. To estimate the $L_2(A_j)$-norm of $|(k, \xi)|_{b, \bar{a}}^{-l}$ we note that for $(k, \xi) \in A_j$ we have $|k| \leq 2^b A^b 2^{jb}$. Since $|\xi|_{\bar{a}} \leq |(k, \xi)|_{b, \bar{a}}$ we apply the transformation of Lemma 1.4.5 with the abbreviation $\psi = (\varphi_1, \ldots, \varphi_{n-1})$ to obtain

$$\left\| |(k,\xi)|_{b,\vec{a}}^{-l} \right\|_{L_2(A_j)} \leq \left(\frac{4^l 2^{-lj}}{B^l} \sum_{|k| \leq (2A)^b 2^{jb}} \int_{B_{\vec{a}}(0, A2^{j+1})} 1 \, d\xi \right)^{\frac{1}{2}}$$

$$\leq c(l, A, B, b) \left(2^{-lj} 2^{jb} \int_0^{A2^{j+1}} \int_{S^{n-1}} r^{|\vec{a}|-1} g(\psi) \, d\psi \, dr \right)^{\frac{1}{2}}.$$

Since g is a bounded function, S^{n-1} a compact set and $|\vec{a}| > 0$ we have

$$\left\| |(k,\xi)|_{b,\vec{a}}^{-l} \right\|_{L_2(A_j)} \lesssim (2^{-lj} 2^{j(|\vec{a}|+b)})^{\frac{1}{2}}.$$

We take $\nu \in \mathbb{N}$ such that $\nu \geq \frac{1}{\tau}$, τ as in (1.13), and derive

$$|(k,\xi)|_{b,\vec{a}}^l \leq \left(n + 1 + |k|^\nu + \sum_{k=1}^n |\xi_k|^\nu \right)^l \leq (n+2)^{l-1} \left[(n+1)^{l\nu} + |k|^{l\nu} + \sum_{k=1}^n |\xi_k|^{l\nu} \right].$$

Thus we have

$$\sum_{\substack{|\alpha| \leq 2n+2m \\ \gamma \leq \alpha}} \left\| |(k,\xi)|_{b,\vec{a}}^l D^{\alpha-\gamma} \mathscr{F}\varphi \right\|_{L_\infty(\widehat{G}_n)} \leq (n+2)^{2l-1} \sum_{|\gamma| \leq 2n+2m} \left[\hat{\rho}_{0,\gamma,0}(\hat{\varphi}) + \hat{\rho}_{0,\gamma,l\nu}(\hat{\varphi}) + \sum_{k=1}^n \hat{\rho}_{l\nu e_k, \gamma, 0}(\hat{\varphi}) \right].$$

If we combine all previous estimates we derive

(1.20) $$|\langle u_j, \varphi \rangle| \lesssim 2^{j(m + \frac{|\vec{a}|+b}{2}) - \frac{1}{2}j} \sum_{|\gamma| \leq 2n+2m} \left[\hat{\rho}_{0,\gamma,0}(\hat{\varphi}) + \hat{\rho}_{0,\gamma,l\nu}(\hat{\varphi}) + \sum_{k=1}^n \hat{\rho}_{l\nu e_k, \gamma, 0}(\hat{\varphi}) \right].$$

Thus by choosing $l > 2(m + |\vec{a}| + b)$ we conclude that $\sum_{j=0}^\infty u_j$ is a Cauchy sequence in $\mathscr{S}'(G_n)$ and hence convergent. □

We continue with a convergence result important for the considerations of Chapter 3. Let the function ψ be an element of $\mathscr{S}(\mathbb{R}^{n+1})$ such that

(1.21) $$\psi(x) = 1 \quad \text{for all } x \in B_{b,\vec{a}}(0, R)$$

for some $R > 0$.

Lemma 1.6.2. Let $u \in \mathscr{S}'(G_n)$ and ψ as in (1.21). The sequence $\{u^m\}_{m \in \mathbb{N}}$ defined for every $m \in \mathbb{N}$ by $u^m := \mathscr{F}^{-1}\left[\psi(2^{-m(b,\vec{a})} \cdot)\big|_{\widehat{G}_n} \mathscr{F}u \right]$ converges to u in $\mathscr{S}'(G_n)$ for $m \to \infty$.

Proof: For $\varphi \in \mathscr{S}(G_n)$ and $\psi^m := \psi(2^{-m(b,\vec{a})} \cdot)\big|_{\widehat{G}_n}$ it holds

$$\langle u^m, \varphi \rangle = \langle \mathscr{F}u, \psi^m \mathscr{F}_{\widehat{G}_n}^{-1} \varphi \rangle$$

Proof: By Theorem 1.5.11 we know that $u \in C^\infty(G_n)$ with at most polynomial growth. Hence if we take $\psi \in C_0^\infty(\mathbb{R}^n)$ such that $\psi(x) = 1$ for all $x \in K$ we have $\mathscr{F}_{\mathbb{R}^n} u = \psi \mathscr{F}_{\mathbb{R}^n} u$ and therefore

$$(1.28) \qquad u(t,x) = \int_{\mathbb{R}^n} \mathscr{F}_{\mathbb{R}^n}^{-1} \psi(x-y) u(t,y) \mathrm{d}y.$$

From this we derive

$$\partial_k u(t, x-z) = \int_{\mathbb{R}^n} \partial_k \mathscr{F}_{\mathbb{R}^n}^{-1} \psi(x-z-y) u(t,y) \mathrm{d}y.$$

By using

$$(1.29) \qquad \frac{1+|x-y|}{1+|z|} \leq \frac{1+|x-y-z|+|z|}{1+|z|} \leq 1+|x-y-z|$$

we obtain

$$\begin{aligned}
\frac{|\partial_k u(t, x-z)|}{(1+|z|)^n} &\leq \int_{\mathbb{R}^n} |\partial_k \mathscr{F}_{\mathbb{R}^n}^{-1} \psi(x-y-z)| |u(t,y)| \frac{(1+|x-y-z|)^n}{(1+|x-y|)^n} \mathrm{d}y \\
&\leq \sup_{w \in \mathbb{R}^n} \frac{|u(t,w)|}{(1+|x-w|)^n} \int_{\mathbb{R}^n} |\partial_k \mathscr{F}_{\mathbb{R}^n}^{-1} \psi(x-y-z)|(1+|x-y-z|)^n \mathrm{d}y \\
&= \sup_{w \in \mathbb{R}^n} \frac{|u(t, x-w)|}{(1+|w|)^n} \int_{\mathbb{R}^n} |\partial_k \mathscr{F}_{\mathbb{R}^n}^{-1} \psi(y)|(1+|y|)^n \mathrm{d}y = c \sup_{w \in \mathbb{R}^n} \frac{|u(t, x-w)|}{(1+|w|)^n}.
\end{aligned}$$

Note that by our construction $\mathscr{F}^{-1}\psi \in \mathscr{S}(\mathbb{R}^n)$ and therefore the integral exists, hence taking the supremum on the left side with respect to $z \in \mathbb{R}^n$ yields the result. Furthermore it is clear that c is independent of M. $\qquad \square$

Under the same assumptions we continue with a similar result.

Lemma 1.7.5. *In the setting of Lemma 1.7.4 and for $0 < \delta \leq 1$ there exists a constant $c > 0$ independent of δ such that*

$$(1.30) \qquad \sup_{z \in \mathbb{R}^n} \frac{|u(t, x-z)|}{(1+|z|)^n} \leq c\delta \sup_{z \in \mathbb{R}^n} \frac{|\nabla u(t, x-z)|}{(1+|z|)^n} + c\delta^{-n} Mu(t,x).$$

Proof: To proof the estimate we first need some auxiliary result. Let $g : \mathbb{R}^n \to \mathbb{C}$ be a continuously differentiable function. By the fundamental theorem of calculus we obtain

$$|g(x)| - |g(y)| \leq \sup_{\xi \in [x,y]} |\nabla g(\xi)| |x-y|$$

If we restrict ourself to the cube $\overline{Q_\delta(0)}$, integrate the inequality with respect to y over $Q_\delta(0)$ and divide by $|Q_\delta(0)| = 2^n \delta^n$ we obtain

$$|g(x)| \leq 2\sqrt{n}\delta \sup_{\xi \in Q_\delta(0)} |\nabla g(\xi)| + 2^{-n}\delta^{-n} \int_{Q_\delta(0)} |g(y)| \mathrm{d}y$$

for every $\delta > 0$ and every $x \in \overline{Q_\delta(0)}$. We apply this result to $u(t, x - w - \cdot)$ for fixed t at 0 to conclude

$$(1.31) \qquad |u(t, x - w)| \le 2\sqrt{n}\delta \sup_{\xi \in Q_\delta(0)} |\nabla u(t, x - w - \xi)| + 2^{-n}\delta^{-n} \int_{Q_\delta(0)} |u(t, x - w - y)| dy.$$

We fix $w \in \mathbb{R}^n$ and since $\delta \le 1$ we have the estimate

$$2^{-n} \int_{Q_\delta(0)} |u(t, x - w - y)| dy \le 2^{-n} \int_{Q_1(0)} |u(t, x - w - y)| dy$$

$$\le 2^{-n} \int_{Q_{1+|w|}(0)} |u(t, x - y)| dy \le (1 + |w|)^n M |u(t, \cdot)|(x).$$

Dividing (1.31) by $(1 + |w|)^n$, applying the previous estimate and the idea of (1.29) yields

$$\frac{|u(t, x - w)|}{(1 + |w|)^n} \le 2\sqrt{n}\delta \sup_{\xi \in Q_\delta(0)} \frac{|\nabla u(t, x - w - \xi)|}{(1 + |w|)^n} + \delta^{-n} M |u(t, \cdot)|(x)$$

$$\le 2\sqrt{n}\delta \sup_{\xi \in Q_\delta(0)} \frac{|\nabla u(t, x - (w + \xi))|(1 + |\xi|)^n}{(1 + |w + \xi|)^n} + \delta^{-n} M |u(t, \cdot)|(x)$$

$$\le 2\sqrt{n}2^n\delta \sup_{z \in \mathbb{R}^n} \frac{|\nabla u(t, x - z)|}{(1 + |z|)^n} + \delta^{-n} M |u(t, \cdot)|(x).$$

Now the right hand side is independent of w and hence the result follows. $\qquad\square$

With the previous results we can prove a similar result as in Johnsen and Sickel [56, Proposition 3.12]; note that the version here is a bit simpler since we only consider the case of $1 < \vec{p}$.

Proposition 1.7.6. *Let $1 < \vec{p}, q < \infty$, $1 < r \le \infty$ and $\{\vec{b^j}\}_{j \in \mathbb{N}_0}$ with $\vec{b^j} \in (0, \infty)^n$ for every $j \in \mathbb{N}_0$. Then there exists a constant $c > 0$ such that*

$$\left\| \{u_j^*(\vec{b^j}; t, x)\}_{j \in \mathbb{N}_0} \right\|_{L_{q, \vec{p}}(G_n; \ell_r)} \le c \|\{u_j\}_{j \in \mathbb{N}_0}\|_{L_{q, \vec{p}}(G_n; \ell_r)}$$

for all sequences $(u_j)_{j \in \mathbb{N}_0} \subset L_{q, \vec{p}}(G_n; \ell_r)$ such that $\operatorname{supp} \mathscr{F} u_j \subset M_j \times Q_{\vec{b^j}}(0)$, where M_j are compact sets in $\mathbb{Z}$.

Proof: To every u_j we define the function $g_j(t, x) := u_j\left(t, \frac{x_1}{b_1^j}, \frac{x_2}{b_2^j}, \ldots, \frac{x_n}{b_n^j}\right)$. It is clear that $\operatorname{supp} \mathscr{F} g_j \subset M_j \times Q_1(0)$ and we therefore have

$$g_j^*(t, x) := \sup_{z \in \mathbb{R}^n} \frac{|g_j(t, x - z)|}{(1 + |z|)^n} \le cM g_j(t, x) \le cM_S g_j(t, x),$$

by combining Lemmata 1.7.4 and 1.7.5 and choosing δ suitably small. The last estimate was already stated in Lemma 1.7.3 and the constant c does not depend on j since all $\mathscr{F} g_j$ have the same support with respect to x. We furthermore derive

$$g_j^*(t, x) = \sup_{z \in \mathbb{R}^n} \frac{\left|u_j\left(t, \frac{x_1 - z_1}{b_1^j}, \ldots, \frac{x_n - z_n}{b_n^j}\right)\right|}{(1 + |z|)^n} = \sup_{z \in \mathbb{R}^n} \frac{\left|u_j\left(t, \frac{x_1}{b_1^j} - z_1, \ldots, \frac{x_n}{b_n^j} - z_n\right)\right|}{(1 + |\vec{b^j}z|)^n}.$$

Combining all previous results yields

$$u_j^*(\vec{b^j};t,x) = g_j^*(t,\vec{b^j}x) \leq c_1[M_S g_j(t,\cdot)](\vec{b^j}x) = c_1[M_S g_j(t,\vec{b_j}\cdot)](x) = c_1[M_S u_j(t,\cdot)](x),$$

where we used that the strong maximal operator commutes with the linear transformation $x \mapsto \gamma x$ for $\gamma \in \mathbb{R}_+^n$, see Definition 1.4.1 for this scaling. Now the result follows from Lemma 1.7.3 with the estimate (1.25). $\qquad \square$

Remark 1.7.7. In the previous proposition the compact support with respect to time is only required to allow us to apply the previous two lemmata and does not influence the constant at all. The requirements of previous results can be weakened if one somehow ensures that u is in fact a function such that equation (1.28) holds for all $t \in \mathbb{T}$.

1.8 The Laplace Operator with Navier Type Boundary Conditions

This section will provide results regarding the Stokes operator with Navier-type boundary conditions on a bounded domain Ω for the study of Chapter 4. The equations are given by

$$(1.32) \qquad \begin{cases} \lambda H - \Delta H = f & \text{in } \Omega, \\ \text{div } H = 0 & \text{in } \Omega, \\ H \cdot n = 0, \quad \text{curl } H \times n = 0 & \text{on } \partial\Omega, \end{cases}$$

where $f \in L_{p,\sigma}(\Omega)$ and Ω is of class $C^{2,1}$. The interesting fact about this partial differential equation is, that the condition regarding the divergence is optional, because $f \in L_{p,\sigma}(\Omega)$ guarantees the solution to be solenoidal, see Al Baba, Amrouche and Escobedo [3, Remark 4.2] or the beginning of Section 4.1 for details. The following result can also be found in the cited work.

Theorem 1.8.1. *Let $\lambda \in \mathbb{C}_*$ be such that $\operatorname{Re}\lambda \geq 0$, $1 < p < \infty$ and $f \in L_{p,\sigma}(\Omega)$. Then there exists a unique solution $H \in W_p^2(\Omega)$ to (1.32) which satisfies the estimates*

$$(1.33) \qquad \|H\|_{L_p(\Omega)} \lesssim \frac{\|f\|_{L_p(\Omega)}}{|\lambda|},$$

$$(1.34) \qquad \|H\|_{W_p^2(\Omega)} \lesssim \frac{1+|\lambda|}{|\lambda|}\|f\|_{L_p(\Omega)},$$

with constants independent of λ.

An important estimate in the theory of partial differential equations is given by Poincaré's inequality, *i.e.*, $\|u\|_p \lesssim \|\nabla u\|_p$ if $1 \leq p < \infty$ and $u \cdot n = 0$ on $\partial\Omega$. When dealing with equations (1.32) a natural question is, whether the same inequality holds for curl instead of ∇. It is known that for a general domain this inequality is not valid, even if one adds the L_p-norm of div u on the right hand side, since there exist non-trivial functions $v \in L_p(\Omega)$ such that curl $v = 0$, div $v = 0$ in Ω and $v \cdot n = 0$ on $\partial\Omega$. But if Ω is simply connected we have the following estimate

$$(1.35) \qquad \|u\|_{L_p(\Omega)} \lesssim \|\nabla u\|_{L_p(\Omega)} \lesssim \|\operatorname{curl} u\|_{L_p(\Omega)} + \|\operatorname{div} u\|_{L_p(\Omega)}$$

for any $u \in W_p^1(\Omega)$ with $u \cdot n = 0$ on $\partial\Omega$. For the second estimate, see Amrouche and Seloula [7, Theorem 3.3], the first is an application of Poincaré's inequality. The additional regularity of the domain allows Theorem 1.8.1 to be extended to $\lambda = 0$.

Proposition 1.8.2. *Let Ω be a simply connected domain and $f \in \mathrm{L}_{p,\sigma}(\Omega)$. Then there exists a unique solution $H \in \mathrm{W}_p^2(\Omega)$ to (1.32) with $\lambda = 0$ which satisfies the estimate*

$$(1.36) \qquad \|H\|_{\mathrm{W}_p^2(\Omega)} \lesssim \|f\|_{\mathrm{L}_p(\Omega)}.$$

A proof of this result can be found in Amrouche and Seloula [6, Proposition 4.7], the additional compatibility conditions are satisfied for a simply connected domain because $\mathbf{K}_T^p(\Omega) = \{0\}$, see Chapter 2 of the cited paper. Similar to (1.9) the trace $u \times n$ can be defined as an element of $\mathrm{W}_p^{-\frac{1}{p}}(\partial\Omega)^3$ for any function $u \in \mathrm{L}_p(\Omega)^3$ with $\operatorname{curl} u \in \mathrm{L}_p(\Omega)^3$ by

$$(1.37) \qquad \langle u \times n, \varphi \rangle_{\partial\Omega} := \int_\Omega u(x) \cdot \operatorname{curl} \varphi(x) \, \mathrm{d}x - \int_\Omega \operatorname{curl} u(x) \cdot \varphi(x) \, \mathrm{d}x$$

for all $\varphi \in \mathrm{W}_{p'}^1(\Omega)^3$. This is again justified by the fact that every u with the above properties can be approximated by elements in $\mathcal{D}(\overline{\Omega})$ and for these the identity holds by an application of the divergence theorem, see Amrouche and Seloula [7, Section 2] for details.

CHAPTER 2

In this thesis we extend the theory of anisotropic function spaces on $\mathbb{R}^n$ to the locally compact abelian group G_n, see Chapter 3, and construct time-periodic solutions to a partial differential equation, see Chapter 4. Naturally, this requires some results in the field of harmonic analysis and, although there has been a lot of research in this field, the nature of the anisotropy and the periodicity modelled by the locally compact group restrict the usage of standard theory. Therefore this chapter will provide proofs of necessary statements and will be quite thorough in the proofs to keep the thesis self-contained.

In the first section we consider Nikol'skij-Plancherel-Polya-type inequalities, $i.e.$, the estimate

$$\|\varphi\|_{L_q(\mathbb{R}^n)} \lesssim R^{n\left(\frac{1}{p}-\frac{1}{q}\right)} \|\varphi\|_{L_p(\mathbb{R}^n)},$$

which holds for all functions $\varphi \in \mathscr{S}(\mathbb{R}^n)$ such that $\operatorname{supp} \mathscr{F}_{\mathbb{R}^n}\varphi \subset B_R(0)$ and $1 \leq p \leq q \leq \infty$, see Triebel [80, Section 1.3.2] for a proof. Some authors refer to them as Bernstein inequalities, see for example Bahouri, Chemin and Danchin [11, Section 2.1]. This kind of estimates has been extended to functions with values in some space ℓ_r for $1 \leq r \leq \infty$, see Triebel [80, Section 2.7]. The generalisation to the case of $L_{\vec{p}}(\mathbb{R}^n)$-spaces can be found in Johnsen and Sickel [55, Section 2] and we build on their result to expand the inequality to $L_{q,\vec{p}}(G_n)$-spaces.

Section 2.2 concerns the extension of the transference principle. It states that if m is a continuous and bounded function on a locally compact abelian group G that is an L_p Fourier multiplier, then for any continuous group homomorphism $\Phi : \widehat{G} \to \widehat{H}$, where H is a locally compact abelian group, the function $m \circ H$ is an L_p Fourier multiplier on H. This tool allows to transfer the $L_p(\mathbb{R}^n)$-multiplier results of Marcinkiewicz or Mihlin-Hörmander, see Grafakos [41, Section 6.2], to the case of $L_p(G_n)$-multipliers. We extend the existing theory to be able to transfer $L_{\vec{p}}(\mathbb{R}^n, \mathcal{B}_1)$-multipliers for $\mathcal{B}_1 \in \{\mathbb{C}, \ell_r\}$ for any $1 \leq r \leq \infty$, which will be of importance for Chapter 3.

The main goal of section 2.3 is to extend the Littlewood-Paley decomposition to an anisotropic decomposition of G_n, $i.e.$, the estimate

$$\|f\|_{L_p(\mathbb{R}^n)} \lesssim \left\| \left(\sum_{j=0}^{\infty} |\mathscr{F}_{\mathbb{R}^n}^{-1} \varphi_j \mathscr{F} f|^2 \right)^{\frac{1}{2}} \right\|_{L_p(\mathbb{R}^n)} \lesssim \|f\|_{L_p(\mathbb{R}^n)},$$

where $1 < p < \infty$ and $\sum_{j=0}^{\infty} \varphi_j = 1$ with $\varphi \in C_0^{\infty}(\mathbb{R}^n)$. For a proof and more details see Grafakos [41, Section 6.1] for example. Littlewood-Paley decompositions on groups such as $\mathbb{T}$ or $\mathbb{Z}$ can for example be found in Edwards and Gaudry [29, Chapter 4 and 7]. They cover combinations of $\mathbb{T}$ and $\mathbb{R}^n$ but only in the isotropic L_p-setting and hence their results do not provide the results needed for this thesis. Lizorkin [65, Theorem 2] proved an isotropic Littlewood-Paley decomposition in $L_{\vec{p}}(\mathbb{R}^n)$-spaces. The proof relies on a Calderón-Zygmund decomposition together with an extension theorem for convolution operators. Such a theorem is standard in the L_p-setting, see Grafakos [41, Section 5.3.3] and in the $L_{\vec{p}}$-setting, see Benedek, Calderón and Panzone [12, Theorem 2]. As all of the previous results are in the isotropic setting we will extend the results to the anisotropic setting in Subsection 2.3.1. Afterwards we are going to apply the previous results to conclude an anisotropic Littlewood-Paley decomposition for $L_{q,\vec{p}}(G_n)$.

2.1 Nikol'skij-Plancherel-Polya Type Inequalities

One of the main tools to extend Nikol'skij-Plancherel-Polya-type inequalities to G_n are L_p-estimates of the Dirichlet kernel D_K. For a fixed $K \in \mathbb{N}$ it is given by

$$(2.1) \qquad D_K(t) = \sum_{|j| \le K} e^{ij\frac{2\pi}{T}t} = \frac{e^{i\frac{2\pi}{T}(K+1)t} - e^{-i\frac{2\pi}{T}Kt}}{e^{i\frac{2\pi}{T}t} - 1} = \frac{e^{i\frac{2\pi}{T}(K+\frac{1}{2})t} - e^{-i\frac{2\pi}{T}(K+\frac{1}{2})t}}{e^{i\frac{\pi}{T}t} - e^{-i\frac{\pi}{T}t}}$$

$$= \frac{\sin\left(\left[\frac{2\pi}{T}K + \frac{\pi}{T}\right]t\right)}{\sin\left(\frac{\pi}{T}t\right)}.$$

One can immediately see that we have $\mathscr{F}_{\mathbb{T}}[D_K(t)] = \chi_{[-K,K]}\big|_{\mathbb{Z}}$, and in Lemma 5.1.3 the estimate

$$(2.2) \qquad \|D_K\|_{L_p(\mathbb{T})} \lesssim K^{1-\frac{1}{p}}$$

will be shown for all $1 < p < \infty$. The case $p = \infty$ can easily be seen as D_K has a maximum at $t = 0$, but it is of no interest for the following proofs. We start with an estimate for a single function.

Lemma 2.1.1. *Let $1 \le \vec{p} \le \vec{r} \le \infty$ and $1 \le q_1 \le q_2 \le \infty$. For every function $f \in L_{q_1,\vec{p}}(G_n)$ with*

$$(2.3) \qquad \operatorname{supp} \mathscr{F}_{G_n} f \subset [-K, K] \times [-R_1\, R_1] \times \cdots \times [-R_n, R_n] \subset \mathbb{R} \times \mathbb{R}^n$$

it holds $f \in L_{q_2,\vec{r}}(G_n)$ together with the estimate

$$(2.4) \qquad \|f\|_{L_{q_2,\vec{r}}(G_n)} \lesssim K^{\frac{1}{q_1}-\frac{1}{q_2}} \prod_{j=1}^{n} R_j^{\frac{1}{p_j}-\frac{1}{r_j}} \|f\|_{L_{q_1,\vec{p}}(G_n)}.$$

Proof: By Theorem 1.5.11 we see that every function that fulfils (2.3) is in $L_1(\mathbb{T})$ for all $x \in \mathbb{R}^n$, which allows to apply the Fourier transform $\mathscr{F}_{\mathbb{T}}$ directly. Furthermore we can assume that $K \in \mathbb{Z}$, as it only matters which elements of $\mathbb{Z}$ are in the set $[-K, K]$. For an arbitrary K we take the largest element in the set, finish the proof and estimate afterwards by K. From (2.3) we obtain the identity $\mathscr{F}_{\mathbb{T}}[f](k, x) = \mathscr{F}_{\mathbb{T}}[f](k, x) \cdot \mathscr{F}_{\mathbb{T}}[D_K](k)$ and applying the inverse Fourier transform yields

$$f(t, x) = (f(\cdot, x) *_{\mathbb{T}} D_K)(t).$$

By applying Minkowski's integral inequality we have

$$\|f\|_{L_{q_2,\vec{r}}(G_n)} = \|(f(\cdot,x) *_{\mathbb{T}} D_K)(t)\|_{L_{q_2,\vec{r}}(G_n)} \leq \left\| \int_{\mathbb{T}} \|f(s,\cdot)\|_{L_{\vec{r}}(\mathbb{R}^n)} |D_K(t-s)| \mathrm{d}s \right\|_{L_{q_2}(\mathbb{T})}.$$

Theorem 1.5.11 implies that for $s \in \mathbb{T}$ the Fourier transform of the function $x \mapsto f(s,x)$ has support in $[-R_1 R_1] \times \cdots \times [-R_n, R_n]$ and hence we can apply the inequality from Johnsen and Sickel [55, Proposition 4] to obtain

$$\left\| \int_{\mathbb{T}} \|f(s,\cdot)\|_{L_{\vec{r}}(\mathbb{R}^n)} |D_K(t-s)| \mathrm{d}s \right\|_{L_{q_2}(\mathbb{T})} \lesssim \prod_{j=1}^{n} R_j^{\frac{1}{p_j}-\frac{1}{r_j}} \left\| \int_{\mathbb{T}} \|f(s,\cdot)\|_{L_{\vec{p}}(\mathbb{R}^n)} |D_K(t-s)| \mathrm{d}s \right\|_{L_{q_2}(\mathbb{T})}.$$

This proves the case for $q_1 = q_2$ as we omit the function D_K and directly apply the previous estimate to the $L_{q_2,\vec{r}}(G_n)$-norm of f. For $q_1 < q_2$ we apply Young's inequality to the convolution integral with $1 + \frac{1}{q_2} = \frac{1}{q_1} + \frac{1}{u}$. Note that we can always find such a u in $(1,\infty)$ as $\frac{1}{q_2} < \frac{1}{q_1}$. Combining the previous ideas yields

$$\|f\|_{L_{q_2,\vec{r}}(G_n)} \lesssim \prod_{j=1}^{n} R_j^{\frac{1}{p_j}-\frac{1}{r_j}} \|f\|_{L_{q_1,\vec{p}}(G_n)} \|D_K\|_{L_u(\mathbb{T})}.$$

We use the estimate (2.2) to derive

$$\|f\|_{L_{q_2,\vec{r}}(G_n)} \lesssim K^{\frac{1}{q_1}-\frac{1}{q_2}} \prod_{j=1}^{n} R_j^{\frac{1}{p_j}-\frac{1}{r_j}} \|f\|_{L_{q_1,\vec{p}}(G_n)}$$

since $1 - \frac{1}{u} = \frac{1}{q_1} - \frac{1}{q_2}$. $\qquad\qquad\square$

As a next step we extend the previous result to the case of sequences of functions $\{f_j\}_{j\in\mathbb{N}_0}$, where the Fourier transform of each entry f_j has compact support scaling with the index $j \in \mathbb{N}_0$.

Theorem 2.1.2. Let $1 \leq \vec{p} \leq \vec{r} < \infty$, $1 \leq q_1 \leq q_2 < \infty$ such that $(q_1,\vec{p}) \neq (q_2,\vec{r})$, and $1 \leq q \leq \infty$. Let $(f_j)_{j\in\mathbb{N}_0}$ be a sequence in $\mathscr{S}'(G_n)$ such that there exist constants $A, B > 0$ with

$$(2.5) \qquad \mathrm{supp}\, \mathscr{F}_{G_n} f_j \subset [-BK^j, BK^j] \times [-AR_1^j AR_1^j] \times \cdots \times [-AR_n^j, AR_n^j] \subset \mathbb{R} \times \mathbb{R}^n$$

for fixed numbers $K, R_1, R_2, \ldots, R_n > 1$. Then the following estimate holds

$$(2.6) \qquad \|\{f_j\}_{j\in\mathbb{N}_0}\|_{L_{q_2,\vec{r}}(G_n,\ell_q)} \lesssim \left\| \sup_{j\in\mathbb{N}_0} K^{\frac{j}{q_1}-\frac{j}{q_2}} \prod_{k=1}^{n} R_k^{\frac{j}{p_k}-\frac{j}{r_k}} |f_j| \right\|_{L_{q_1,\vec{p}}(G_n)}.$$

Proof: The proof is done by a succession of estimates, meaning we show the estimate of one component of $(q_2,\vec{r})$ at a time. Hence we are left to show that the inequality holds, if only one component of $(q_2,\vec{r})$ changes. Without loss of generality we can assume $q_1 < q_2$, as it does not matter if we integrate over $\mathbb{T}$ or $\mathbb{R}$ in the proof.

The monotonicity of the ℓ_q-spaces allows to assume $q = 1$ and hence by applying Minkowski's integral inequality we have

$$\|\{f_j\}_{j\in\mathbb{N}_0}\|_{L_{q_2,\vec{r}}(G_n,\ell_q)}^{q_2} \leq \int_{\mathbb{T}} \left(\sum_{j=0}^{\infty} \|f_j(t,\cdot)\|_{\vec{r}} \right)^{q_2} \mathrm{d}t.$$

As M was arbitrary, considering $M \to \infty$ yields the result. □

To define the class of multipliers which can be transferred we introduce the following function. Let $\chi_n : \mathbb{R} \to \mathbb{R}$ be the characteristic function of $[-n, n]$ and define

$$(2.9) \qquad \phi_n(x) = \frac{\chi_n * \chi_n(x)}{\sqrt{2\pi}2n} = \frac{1}{\sqrt{2\pi}2n} \int_{-n}^{n} \chi_n(x - y) \, \mathrm{d}y = \frac{1}{\sqrt{2\pi}} \begin{cases} 1 - \frac{|x|}{2n} & \text{for } |x| \le 2n, \\ 0 & \text{for } |x| > 2n. \end{cases}$$

Furthermore it is easy to see that

a) $\hat{\phi}_n(\xi) = \sqrt{2\pi}\frac{\hat{\chi}_n^2(\xi)}{\sqrt{2\pi}2n} = \frac{\sin^2(n\xi)}{n\pi\xi^2} \ge 0$.

b) $\int_{\mathbb{R}} \hat{\phi}_n(\xi) \, \mathrm{d}\xi = \int_{\mathbb{R}} \frac{\hat{\chi}_n^2(\xi)}{2n} \, \mathrm{d}\xi = \int_{\mathbb{R}} \frac{\chi_n^2(\xi)}{2n} \, \mathrm{d}\xi = 1$.

c) It holds $\lim_{n\to\infty} \int_{A^c} \hat{\phi}_n(\xi) \, \mathrm{d}\xi = 0$ for every compact set $A \subset \mathbb{R}$ containing the origin in its interior.

The first result is a straightforward calculation, the second follows from Plancherel's Theorem and the third from the fact that $\xi \mapsto \frac{1}{\xi^2}$ is integrable on A^c. Note that these properties also imply that $\hat{\phi}_n$ is an approximate identity.

Definition 2.2.2. A bounded measurable function $m : \mathbb{R} \times \mathbb{R}^n \to \mathbb{C}$ is *normalized* with respect to $\{\hat{\psi}_n(\xi)\}_{n\in\mathbb{N}} := \{\hat{\phi}_n(\xi_1) \cdot \hat{\phi}_n(\xi_2) \cdots \hat{\phi}_n(\xi_{n+1})\}_{n\in\mathbb{N}}$ if $\lim_{n\to\infty}(m * \hat{\psi}_n)(t, x) = m(t, x)$ for all $(t, x) \in \mathbb{R} \times \mathbb{R}^n$.

This guarantees that the values $m(k, x)$ for $k \in \mathbb{Z}$ are well-defined, as they are limits of continuous uniformly bounded functions. An example of a normalized function is a bounded continuous function, see Grafakos [41, Theorem 1.2.19]. The following Lemma provides the important approximation result for this section, as it constructs compactly supported kernels in $L_1(\mathbb{R}^{n+1})$ that converge to given multipliers without exceeding their norm estimate. Before we state the lemma we recall the convention that all $f_j \in \mathscr{S}(\mathbb{R}^{n+1})$ respectively $f_j \in \mathscr{S}(G_n)$ for all $j \in \mathbb{N}$.

Lemma 2.2.3. *Let $m_j : \mathbb{R}^{n+1} \to \mathbb{R}$ be a sequence of normalized functions which satisfy the estimate*

$$\left\| \left\{ \mathscr{F}_{\mathbb{R}^{n+1}}^{-1}[m_j(\xi)\hat{f}_j(\xi)] \right\}_{j\in\mathbb{N}} \right\|_{L_{\vec{p}}(\mathbb{R}^{n+1};\mathcal{B}_2)} \le c(m)\|f\|_{L_{\vec{p}}(\mathbb{R}^{n+1};\mathcal{B}_1)}.$$

Then there exists a sequence of functions $k_j^l \in L_1(\mathbb{R}^{n+1})$ that converges to m_j for $l \to \infty$ in every point $\xi \in \mathbb{R}^{n+1}$ and all $j \in \mathbb{N}$. Additionally, they satisfy the estimate

$$(2.10) \qquad \left\| \left\{ \mathscr{F}_{\mathbb{R}^{n+1}}^{-1}[\hat{k}_j^l(\xi)\hat{f}_j(\xi)] \right\}_{j\in\mathbb{N}} \right\|_{L_{\vec{p}}(\mathbb{R}^{n+1};\mathcal{B}_2)} \le c(m)\|f\|_{L_{\vec{p}}(\mathbb{R}^{n+1};\mathcal{B}_1)}$$

for all $l \in \mathbb{N}$, and for fixed $l \in \mathbb{N}$ the functions k_j^l have uniform compact support with respect to $j \in \mathbb{N}$.

Proof: We define

$$m_j^l(\xi) = (\hat{\psi}_l * m_j)(\xi)$$

and by the assumptions that all m_j are normalized, we conclude $m_j^l(\xi) \to m_j(\xi)$ for $l \to \infty$ for all $j \in \mathbb{N}$ and all $\xi \in \mathbb{R}^{n+1}$. Furthermore, the estimate

$$(2.11) \qquad \|m_j^l\|_{L_\infty(\mathbb{R}^{n+1})} \le \|m_j\|_{L_\infty(\mathbb{R}^{n+1})}\|\hat{\psi}_l\|_{L_1(\mathbb{R}^{n+1})} = \|m_j\|_{L_\infty(\mathbb{R}^{n+1})}$$

holds uniformly in $l \in \mathbb{N}$.

As a next step we show that the functions m_j^l fulfil (2.10) with a constant bounded by $c(m)$ uniformly in $l \in \mathbb{N}$. With $c_{n+1} = (2\pi)^{-\frac{n+1}{2}}$ it holds

$$\left\| \mathscr{F}_{\mathbb{R}^{n+1}}^{-1}[m_j^l(\xi)\hat{f}_j(\xi)] \right\|_{L_{\vec{p}}(\mathbb{R}^{n+1};\mathcal{B}_2)} = \left\| c_{n+1} \int_{\mathbb{R}^{n+1}} \int_{\mathbb{R}^{n+1}} m_j(\xi-y)\hat{\psi}_l(y)\,dy\hat{f}_j(\xi)e^{i(\cdot)\cdot\xi}\,d\xi \right\|_{L_{\vec{p}}(\mathbb{R}^{n+1};\mathcal{B}_2)}$$

$$= \left\| c_{n+1} \int_{\mathbb{R}^{n+1}} \int_{\mathbb{R}^{n+1}} m_j(\xi-y)\hat{\psi}_l(y)\hat{f}_j(\xi)e^{i(\cdot)\cdot\xi}\,d\xi\,dy \right\|_{L_{\vec{p}}(\mathbb{R}^{n+1};\mathcal{B}_2)}$$

$$\leq \int_{\mathbb{R}^{n+1}} \left\| c_{n+1} \int_{\mathbb{R}^{n+1}} m_j(\xi-y)\hat{\psi}_l(y)\hat{f}_j(\xi)e^{i(\cdot)\cdot\xi}\,d\xi \right\|_{L_{\vec{p}}(\mathbb{R}^{n+1};\mathcal{B}_2)} dy$$

$$\leq \int_{\mathbb{R}^{n+1}} \left\| c_{n+1} \int_{\mathbb{R}^{n+1}} m_j(z)\hat{\psi}_l(y)\hat{f}_j(z+y)e^{i(\cdot)\cdot(z+y)}\,dz \right\|_{L_{\vec{p}}(\mathbb{R}^{n+1};\mathcal{B}_2)} dy.$$

Since all integrals are well-defined and exist by the regularity assumptions made on f_j we can interchange the order of integration, and by applying Minkowski's integral inequality the $L_{\vec{p}}(\mathbb{R}^{n+1};\mathcal{B}_2)$-norm moves inside the integral; in the case of $r = \infty$ moving the supremum into the integral satisfies the same estimate. For the next estimate we note that the term $e^{i(\cdot)y}$ does not change the integral with respect to z and has no influence on the norm and $\mathscr{F}_{\mathbb{R}^{n+1}}[e^{-i(\cdot)\cdot y}f(\cdot)] = \mathscr{F}_{\mathbb{R}^{n+1}}[f](y+\xi)$. Since the functions $\hat{\phi}_l$ have integral 1 for all $l \in \mathbb{N}$ we obtain

$$\left\| \mathscr{F}_{\mathbb{R}^{n+1}}^{-1}[m_j^l(\xi)\hat{f}(\xi)] \right\|_{L_{\vec{p}}(\mathbb{R}^{n+1};\mathcal{B}_2)} \leq \int_{\mathbb{R}^{n+1}} \hat{\psi}_l(y) \left\| \mathscr{F}_{\mathbb{R}^{n+1}}^{-1}\left[m_j(\xi)\mathscr{F}_{\mathbb{R}^{n+1}}[e^{-i(\cdot)\cdot y}f_j(\cdot)](\xi)\right] \right\|_{L_{\vec{p}}(\mathbb{R}^{n+1};\mathcal{B}_2)} dy$$

$$(2.12) \qquad\qquad \leq c(m) \int_{\mathbb{R}^{n+1}} \hat{\psi}_l(y) \left\| e^{-iy\cdot(\cdot)}f(\cdot) \right\|_{L_{\vec{p}}(\mathbb{R}^{n+1};\mathcal{B}_1)} dy$$

$$= c(m) \|f\|_{L_{\vec{p}}(\mathbb{R}^{n+1};\mathcal{B}_1)}.$$

The functions m_j^l do not necessarily have compact support, so we take a non-negative function $h \in C_0^\infty(\mathbb{R}^{n+1})$ with integral equal to 1. By defining $h_l(x) = l^{n+1}h(lx)$ we conclude $\hat{h}_l(x) = \hat{h}(l^{-1}\xi) \in \mathscr{S}(\mathbb{R}^{n+1})$. As $\hat{h}(0) = 1$ we derive $\lim_{l\to\infty} \hat{h}_l(\xi) = 1$ for all $\xi \in \mathbb{R}^{n+1}$. Defining the kernels

$$k_j^l = \mathscr{F}_{\mathbb{R}^{n+1}}^{-1}[\hat{h}_l m_j^l]$$

yields $k_j^l \in L_2(\mathbb{R}^{n+1})$ because $m_j^l \in L_\infty(\mathbb{R}^{n+1})$ by (2.11). Furthermore $\hat{k}_j^l(\xi) = \hat{h}_l(\xi)m_j^l(\xi) \to m_j(\xi)$ for $l \to \infty$ and all $\xi \in \mathbb{R}^{n+1}$. As $\hat{h}_l$ are uniformly bounded and by using (2.11) we obtain

$$(2.13) \qquad\qquad \|\hat{k}_j^l(\xi)\|_{L_\infty(\mathbb{R}^{n+1})} = \|\hat{h}_l m_j^l\|_{L_\infty(\mathbb{R}^{n+1})} \leq c\|m_j\|_{L_\infty(\mathbb{R}^{n+1})}.$$

Next we are going to show that the functions k_j^l have uniform compact support in $j \in \mathbb{N}$ for fixed $l \in \mathbb{N}$ as this implies $k_j^l \in L_1(\mathbb{R}^{n+1})$. To avoid dealing with distributional convolutions and Fourier transforms we consider the functions $m_{j,N} := m_j\chi_{B_N(0)} \in L_1(\mathbb{R}^{n+1})$ because m_j is in $L_\infty(\mathbb{R}^{n+1})$. We define the functions

$$k_{j,N}^l := h_l * (\psi_l \cdot \mathscr{F}^{-1}[m_{j,N}])$$

and note that their support is compact, since it is contained in $\operatorname{supp} h_l + \operatorname{supp} \psi_l = \operatorname{supp} h_l + [-2l, 2l]^{n+1}$ thus independent of j and N for every $l \in \mathbb{N}$. We will show $\lim_{N\to\infty} \|k_{j,N}^l -$

$k_j^l\|_{\mathrm{L}_2(\mathbb{R}^{n+1})} = 0$ which implies the compact support of k_j^l. It holds

$$
\begin{aligned}
\lim_{N \to \infty} \|k_{j,N}^l - k_j^l\|_{\mathrm{L}_2(\mathbb{R}^{n+1})} &= \lim_{N \to \infty} \left\| \hat{h}_l m_j^l - \hat{h}_l \cdot [\hat{\psi}_l * m_{j,N}] \right\|_{\mathrm{L}_2(\mathbb{R}^{n+1})} \\
&= \lim_{N \to \infty} \left\| \hat{h}_l(\cdot) \int_{B_N(0)^c} \hat{\psi}_l(\cdot - y) m_j(y) \, \mathrm{d}y \right\|_{\mathrm{L}_2(\mathbb{R}^{n+1})} \\
&= \left\| \hat{h}_l(\cdot) \lim_{N \to \infty} \int_{\mathbb{R}^{n+1}} \hat{\psi}_l(\cdot - y) \chi_{B_N(0)^c}(y) m_j(y) \, \mathrm{d}y \right\|_{\mathrm{L}_2(\mathbb{R}^{n+1})} = 0.
\end{aligned}
$$

Applying the ideas of (2.11) we see that the convolution is a bounded function uniformly in N and hence by Lebesgue's theorem of dominated convergence we can interchange integration and limit. The integral inside of the norm converges since it exists by the same estimate. Therefore uniform compact support in $j \in \mathbb{N}$ for k_j^l holds for each $l \in \mathbb{N}$, thus yielding $k_j^l \in \mathrm{L}_1(\mathbb{R}^{n+1})$. We are left to show that the estimate (2.10) holds. We obtain

$$
\begin{aligned}
\left\| \mathscr{F}_{\mathbb{R}^{n+1}}^{-1}[\hat{k}_j^l(\xi)\hat{f}(\xi)](\cdot) \right\|_{\mathrm{L}_{\vec{p}}(\mathbb{R}^{n+1};\mathcal{B}_2)} &= c_{n+1} \left\| \left(h_l * \mathscr{F}_{\mathbb{R}^{n+1}}^{-1}[m_j^l(\xi)\hat{f}(\xi)] \right)(\cdot) \right\|_{\mathrm{L}_{\vec{p}}(\mathbb{R}^{n+1};\mathcal{B}_2)} \\
&= c_{n+1} \left\| \int_{\mathbb{R}^{n+1}} h_l(y) \mathscr{F}_{\mathbb{R}^{n+1}}^{-1}[m_j^l(\xi)\hat{f}(\xi)](\cdot - y) \, \mathrm{d}y \right\|_{\mathrm{L}_{\vec{p}}(\mathbb{R}^{n+1};\mathcal{B}_2)} \\
&\leq c_{n+1} \left\| \int_{\mathbb{R}^{n+1}} h_l(y) \| \mathscr{F}_{\mathbb{R}^{n+1}}^{-1}[m_j^l(\xi)\hat{f}(\xi)](\cdot - y) \|_{\mathcal{B}_2} \, \mathrm{d}y \right\|_{\mathrm{L}_{\vec{p}}(\mathbb{R}^{n+1})} \\
&\leq c_{n+1} \|h_l\|_{\mathrm{L}_1(\mathbb{R}^{n+1})} \left\| \mathscr{F}_{\mathbb{R}^{n+1}}^{-1}[m_j^l(\xi)\hat{f}(\xi)](\cdot) \right\|_{\mathrm{L}_{\vec{p}}(\mathbb{R}^{n+1};\mathcal{B}_2)} \\
&\leq c_{n+1} c(m) \|f\|_{\mathrm{L}_{\vec{p}}(\mathbb{R}^{n+1};\mathcal{B}_1)} .
\end{aligned}
$$

By the same argument as before we moved the $\mathcal{B}_2$-norm into the integral and applied Young's inequality together with the estimate from (2.12) to conclude the result. $\qquad \square$

To transfer the multiplier to the group we need to calculate the Fourier transform of the operators $\tilde{T}_k$. By the result of Corollary 1.3.5 we can restrict the consideration to functions of the form

$$
(2.14) \qquad \psi_N(t, x) = \sum_{|m| \leq N} \varphi_m(x) e^{i \frac{2\pi}{\mathcal{T}} mt},
$$

where $\varphi_j \in \mathrm{C}_0^\infty(\mathbb{R}^n)$.

Lemma 2.2.4. *Let $h = (h_j)_{j \in \mathbb{N}}$ with $h_j \in \mathrm{L}_1(\mathbb{R}^{n+1})$ for all $j \in \mathbb{N}$ and $\tilde{T}_h$ given by (2.8). Then*

$$
\mathscr{F}_{G_n}[(\tilde{T}_h \psi_N)_j](k, \xi) = (2\pi)^{\frac{n+1}{2}} \hat{h}_j \left(\frac{2\pi}{\mathcal{T}} k, \xi \right) \mathscr{F}[\psi_N](k, \xi)
$$

holds for all $(k, \xi) \in \widehat{G}_n$, every $j \in \mathbb{N}$, and for every function ψ_N of the form as in equation (2.14).

Proof: We have

$$
\mathscr{F}[\psi_N](k, \xi) : \mathbb{Z} \times \mathbb{R}^n \to \mathbb{C} \text{ with } \mathscr{F}[\psi_N](k, \xi) = \begin{cases} \hat{\varphi}_k(\xi) & \text{for } |k| \leq N, \\ 0 & \text{for } |k| > N. \end{cases}
$$

Furthermore it holds

$$(\tilde{T}_h(\psi_N))_j(t,x) = \sum_{|m|\leq N} e^{i\frac{2\pi}{\mathcal{T}}mt} \int_{\mathbb{R}^n} \varphi_m(x-y) \int_{\mathbb{R}} h_j(s,y) e^{-i\frac{2\pi}{\mathcal{T}}ms}\, ds\, dy$$

$$= \sum_{|m|\leq N} e^{i\frac{2\pi}{\mathcal{T}}mt}\varphi_m(x) *_{\mathbb{R}^n} \left[\sqrt{2\pi}\mathscr{F}_{\mathbb{R}} h_j(\cdot,y)\right]\left(\frac{2\pi}{\mathcal{T}}m\right).$$

So by applying $\mathscr{F}$ we see

$$\mathscr{F}\left[(\tilde{T}_h(\psi_N))_j\right](k,\xi) = \mathscr{F}\left[\sum_{|m|\leq N} e^{i\frac{2\pi}{\mathcal{T}}mt}\varphi_m *_{\mathbb{R}^n}\left[\sqrt{2\pi}\mathscr{F}_{\mathbb{R}}h_j(\cdot,y)\right]\left(\frac{2\pi}{\mathcal{T}}m\right)\right]$$

$$= (2\pi)^{\frac{n+1}{2}}\mathscr{F}_{\mathbb{T}}\left(\sum_{|m|\leq N} e^{i\frac{2\pi}{\mathcal{T}}mt}\hat{\varphi}_m(\xi)\hat{h}_j\left(\frac{2\pi}{\mathcal{T}}m,\xi\right)\right)$$

$$= (2\pi)^{\frac{n+1}{2}}\mathscr{F}_{\mathbb{T}}\left(\mathscr{F}_{\mathbb{T}}^{-1}\left(\hat{\psi}_N(\cdot,\xi)\hat{h}_j\left(\frac{2\pi}{\mathcal{T}}\cdot,\xi\right)\right)\right) = (2\pi)^{\frac{n+1}{2}}\hat{\psi}_N(k,\xi)\hat{h}_j\left(\frac{2\pi}{\mathcal{T}}k,\xi\right),$$

hence we obtain the results. $\qquad\square$

This Lemma allows the combinations of all previous results and we are therefore able to prove the transference principle for multipliers from $\mathbb{R}\times\mathbb{R}^n$ to $\widehat{G}_n$, the main theorem of this section.

Theorem 2.2.5. *Let $m_j : \mathbb{R}\times\mathbb{R}^n \to \mathbb{C}$ be a sequence of normalized functions which satisfy the estimate*

$$\left\|\{\mathscr{F}_{\mathbb{R}^{n+1}}^{-1}[m_j(\xi)\hat{f}(\xi)]\}_{j\in\mathbb{N}}\right\|_{\mathrm{L}_{\vec{p}}(\mathbb{R}^{n+1};\mathcal{B}_2)} \leq c(m)\|f\|_{\mathrm{L}_{\vec{p}}(\mathbb{R}^{n+1};\mathcal{B}_1)}$$

for $1\leq \vec{p} < \infty$. Then the functions $m_j : \widehat{G}_n \to \mathbb{C}$ are well-defined and satisfy the estimate

$$(2.15) \qquad \left\|\{\mathscr{F}_{\mathbb{T}\times\mathbb{R}^n}^{-1}[m_j(k,\xi)\hat{f}(k,\xi)]\}_{j\in\mathbb{N}}\right\|_{\mathrm{L}_{\vec{p}}(\mathbb{T}\times\mathbb{R}^n;\mathcal{B}_2)} \leq c(m)\|f\|_{\mathrm{L}_{\vec{p}}(\mathbb{T}\times\mathbb{R}^n;\mathcal{B}_1)}$$

for all $f \in \mathrm{L}_{\vec{p}}(G_n;\mathcal{B}_1)$ such that $f_j \in \mathscr{S}(G_n)$ for all $j\in\mathbb{N}$.

Proof: We find compactly supported kernels $k_j^l \in \mathrm{L}_1(\mathbb{R}^{n+1})$ by Lemma 2.2.3 such that $\hat{k}_j^l(\xi) \to m_j(\xi)$ for all $\xi\in\mathbb{R}^{n+1}$ and $j\in\mathbb{N}$. Hence the restriction of m_j to $\widehat{G}_n$ is well-defined as stated before. Let f be of the form stated in Corollary 1.3.5 and hence each f_j is of the form (2.14). Since the kernels $\hat{k}_j^l$ are uniformly bounded by $\|m_j\|_{\mathrm{L}_\infty(\mathbb{R}^{n+1})}$ by inequality (2.13) we derive

$$\mathscr{F}_{G_n}^{-1}[m_j(k,\xi)\hat{f}_j(k,\xi)] = \mathscr{F}_{G_n}^{-1}[\lim_{l\to\infty}\hat{k}_j^l(k,\xi)\hat{f}_j(k,\xi)] = \lim_{l\to\infty}\mathscr{F}_{G_n}^{-1}[\hat{k}_j^l(k,\xi)\hat{f}_j(k,\xi)].$$

Applying the result of Lemma 2.2.4 and defining $\omega = \frac{2\pi}{\mathcal{T}}$ yields

$$\left\|\mathscr{F}_{G_n}^{-1}[m_j(k,\xi)\hat{f}(k,\xi)](\cdot)\right\|_{\mathrm{L}_{\vec{p}}(G_n;\mathcal{B}_2)} = \left\|\liminf_{l\to\infty}\mathscr{F}_{G_n}^{-1}[\hat{k}_j^l(\omega k,\xi)\hat{f}(k,\xi)](\cdot)\right\|_{\mathrm{L}_{\vec{p}}(G_n;\mathcal{B}_2)}$$

$$\leq \left\|\liminf_{l\to\infty}\|\mathscr{F}_{G_n}^{-1}[\hat{k}_j^l(\omega k,\xi)\hat{f}(k,\xi)](\cdot)\|_{\mathcal{B}_2}\right\|_{\mathrm{L}_{\vec{p}}(G_n)}$$

$$\leq \liminf_{l\to\infty}\left\|\mathscr{F}_{G_n}^{-1}[\hat{k}_j^l(\omega k,\xi)\hat{f}(k,\xi)](\cdot)\right\|_{\mathrm{L}_{\vec{p}}(G_n;\mathcal{B}_2)}$$

$$\leq \liminf_{l\to\infty}\left\|\tilde{T}_{k_j^l}f\right\|_{\mathrm{L}_{\vec{p}}(G_n;\mathcal{B}_2)} \leq c(m)\|f\|_{\mathrm{L}_{\vec{p}}(G_n;\mathcal{B}_1)}.$$

and 0 otherwise. Set

$$g(x) := \sum_{j=1}^{\infty} \left(\frac{1}{\mu(B_j)} \int_{B_j} f(y)h_j(y) \, dy \right) \chi_{B_j}(x) + f(x)\chi_{\mathcal{O}^c},$$

$$b_j(x) := f(x)h_j(x) - \left(\frac{1}{\mu(B_j)} \int_{B_j} f(y)h_j(x) \, dy \right) \chi_{B_j}(x).$$

All assertions now follow from direct computation and using the properties of Lemma 2.3.2. □

Remark 2.3.4. An estimate that is crucial for further proofs but generally not stated in the decomposition result is

(2.17) $$\|g\|_{L_p(\mathbb{R}^n, \mathcal{B})} \leq \|g\|_{L_1(\mathbb{R}^n, \mathcal{B})}^{\frac{1}{p}} \|g\|_{L_\infty(\mathbb{R}^n, \mathcal{B})}^{\frac{1}{p'}} \leq \|g\|_{L_1(\mathbb{R}^n, \mathcal{B})}^{\frac{1}{p}} C^{\frac{1}{p'}} \alpha^{\frac{1}{p'}},$$

which follows directly from the interpolation inequality of L_p-spaces.

One of the important implications of the previous decomposition is the following extension result. The version presented here can be extended to allow for singular integral kernels, see Grafakos [41, Sections 5.3 and 5.6] whose approach to the proof we will follow. However, since the main goal of this section is to extend the result to $L_{\vec{p}}$ we will refrain from singular integral kernels as they impose problems to the extension. We will give remarks on this problem later on.

Lemma 2.3.5. *Let $\mathcal{B}_1$ and $\mathcal{B}_2$ be Banach spaces and $K : \mathbb{R}^n \to \mathcal{L}(\mathcal{B}_1, \mathcal{B}_2)$ locally integrable. Furthermore we require that the operator*

$$Tf(x) = \int_{\mathbb{R}^n} K(x - y)f(y)dy$$

is bounded from $L_q(\mathbb{R}^n, \mathcal{B}_1)$ to $L_q(\mathbb{R}^n, \mathcal{B}_2)$ for one $q \in (1, \infty]$ with norm B and we assume that K fulfils an anisotropic Hörmander's condition

(2.18) $$\sup_{y \neq 0} \int_{|x|_{\vec{a}} \geq \eta|y|_{\vec{a}}} \|K(x - y) - K(x)\|_{\mathcal{L}(\mathcal{B}_1, \mathcal{B}_2)} \, dx \leq A$$

for some $\eta \in (1, \infty)$ and $A > 0$. Then the operator T extends to a continuous operator from $L_1(\mathbb{R}^n, \mathcal{B}_1)$ to $L_{1,\infty}(\mathbb{R}^n, \mathcal{B}_2)$, i.e. there exists a dimensional constant $C_{n,\vec{a}} > 0$ such that

(2.19) $$\|Tf\|_{L_{1,\infty}(\mathbb{R}^n, \mathcal{B}_2)} \leq C_{n,\vec{a}}(A + B)\|f\|_{L_1(\mathbb{R}^n, \mathcal{B}_1)}$$

for all $F \in L_1(\mathbb{R}^n, B_2)$, and to a bounded operator from $L_p(\mathbb{R}^n, \mathcal{B}_1)$ to $L_p(\mathbb{R}^n, \mathcal{B}_2)$ for all $p \in (1, \infty)$, i.e. there exists a constant $C_{n,p}$ such that

(2.20) $$\|Tf\|_{L_p(\mathbb{R}^n, \mathcal{B}_2)} \leq C_{n,p}(A + B)\|f\|_{L_p(\mathbb{R}^n, \mathcal{B}_1)}$$

for all $f \in L_p(\mathbb{R}^n, \mathcal{B}_1)$.

Proof: We start by showing the $L_{1,\infty}$ estimate, hence let $\beta > 0$ be given. We take $f \in L_1(\mathbb{R}^n, \mathcal{B}_1)$ with compact support and corresponding decomposition $f = g + b$ to $\alpha = \frac{\beta}{2CB}$ from Lemma 2.3.3, here C denotes the constant in the estimates of said lemma. By the following standard inequality

$$\left|\left\{x \in \mathbb{R}^n \mid \|Tf(x)\|_{\mathcal{B}_2} > \beta\right\}\right| \leq \left|\left\{x \in \mathbb{R}^n \mid \|Tg(x)\|_{\mathcal{B}_2} > \frac{\beta}{2}\right\}\right| + \left|\left\{x \in \mathbb{R}^n \mid \|Tb(x)\|_{\mathcal{B}_2} > \frac{\beta}{2}\right\}\right|$$

we can consider g and b separately. For $q < \infty$ we obtain

$$\left|\left\{x \in \mathbb{R}^n \mid \|Tg\|_{\mathcal{B}_2} > \frac{\beta}{2}\right\}\right| \leq \left(\frac{2}{\beta}\right)^q \int_{\mathbb{R}^n} \|Tg(x)\|_{\mathcal{B}_2}^q \, \mathrm{d}x \leq \left(\frac{2B}{\beta}\right)^q \|g\|_{L_q(\mathbb{R}^n, \mathcal{B}_1)}^q$$

$$\leq \left(\frac{2B}{\beta}\right)^q \|g\|_{L_1(\mathbb{R}^n, \mathcal{B}_1)} C^{\frac{q}{q'}} \alpha^{\frac{q}{q'}} = \frac{(2B)^q C^{q-1} \alpha^q}{\beta^q \alpha} \|g\|_{L_1(\mathbb{R}^n, \mathcal{B}_1)}$$

$$= 2B \frac{\|g\|_{L_1(\mathbb{R}^n, \mathcal{B}_1)}}{\beta} \leq 2B \frac{\|f\|_{L_1(\mathbb{R}^n, \mathcal{B}_1)}}{\beta}.$$

For $q = \infty$ we have $\|Tg\|_{L_\infty(\mathbb{R}^n, \mathcal{B}_2)} \leq B \|g\|_{L_\infty(\mathbb{R}^n, \mathcal{B}_1)} \leq CB\alpha = \frac{\beta}{2}$, hence the above measure is 0. Next up we consider Tb and define
$$B_j^* = B_{\bar{a}}(x_j, \eta R_j),$$
which are stretched versions of the balls B_j from Lemma 2.3.3 with the same center. It holds

$$\left|\left\{x \in \mathbb{R}^n \mid \|Tb\|_{\mathcal{B}_2} > \frac{\beta}{2}\right\}\right| \leq \left|\bigcup_j B_j^*\right| + \left|\left\{x \notin \bigcup_j B_j^* \mid \|Tb\|_{\mathcal{B}_2} > \frac{\beta}{2}\right\}\right|$$

$$\leq \eta^{|\bar{a}|} \sum_j |B_j| + \frac{2}{\beta} \int_{(\cup B_j^*)^c} \|Tb(x)\|_{\mathcal{B}_2} \, \mathrm{d}x$$

$$\leq \eta^{|\bar{a}|} 2C^2 B \frac{\|f\|_{L_1(\mathbb{R}^n, \mathcal{B}_1)}}{\beta} + \sum_k \frac{2}{\beta} \int_{(B_j^*)^c} \|Tb_j(x)\|_{\mathcal{B}_2} \, \mathrm{d}x$$

since $b = \sum_j b_j$. Recall that x_j is the middle point of B_j and by construction of B_j^* as well. Since b_j is supported on B_j and has mean value zero we conclude

$$\int_{(B_j^*)^c} \|Tb_j(x)\|_{\mathcal{B}_2} \, \mathrm{d}y = \int_{(B_j^*)^c} \left\| \int_{B_j} K(x-y)b_j(y) \, \mathrm{d}y \right\|_{\mathcal{B}_2} \, \mathrm{d}x$$

$$= \int_{(B_j^*)^c} \left\| \int_{B_j} [K(x-y) - K(x-x_j)]b_j(y) \, \mathrm{d}y \right\|_{\mathcal{B}_2} \, \mathrm{d}x$$

$$\leq \int_{(B_j^*)^c} \int_{B_j} \left\| [K(x-y) - K(x-x_j)]b_j(y) \right\|_{\mathcal{B}_2} \, \mathrm{d}y \, \mathrm{d}x$$

$$\leq \int_{B_j} \|b_j(y)\|_{\mathcal{B}_1} \int_{(B_j^*)^c} \left\| K(x-y) - K(x-x_j) \right\|_{\mathcal{L}(\mathcal{B}_1, \mathcal{B}_2)} \, \mathrm{d}x \, \mathrm{d}y.$$

Furthermore we obtain the following identity

$$\int_{B_j} \|b_j(y)\|_{\mathcal{B}_1} \int_{(B_j^*)^c} \left\| K(x-y) - K(x-x_j) \right\|_{\mathcal{L}(\mathcal{B}_1, \mathcal{B}_2)} \, \mathrm{d}x \, \mathrm{d}y$$

$$= \int_{B_j} \|b_j(y)\|_{\mathcal{B}_1} \int_{(B_j^*)^c} \left\| K((x-x_j) - (y-x_j)) - K(x-x_j) \right\|_{\mathcal{L}(\mathcal{B}_1, \mathcal{B}_2)} \, \mathrm{d}x \, \mathrm{d}y.$$

So for the integral concerning the kernel K we can assume that B_j and B_j^* are centered around the origin. By construction we have $y \in B_j = B_{\vec{a}}(x_j, R_j)$ if and only if $|y|_{\vec{a}} \leq R_j$ and $x \in (B_j^*)^c$ if and only if $|x|_{\vec{a}} \geq \eta R_j$. This enables the usage of (2.18) to the last integral and hence by using the properties of Lemma 2.3.3 we derive

$$\sum_j \int_{(B_j^*)^c} \|Tb_j(y)\|_{\mathcal{B}_2} \, dy \leq \sum_j A\|b_j\|_{L_1(\mathbb{R}^n, \mathcal{B}_1)} \leq AC^2\|f\|_{L_1(\mathbb{R}^n, \mathcal{B}_1)}.$$

Combining all previous estimates yields

$$\left|\{x \in \mathbb{R}^n \mid \|Tf\|_{\mathcal{B}_2} > \beta\}\right| \leq \left[2B + \eta^{|\vec{a}|}2C^2B + 2AC^2\right]\frac{\|f\|_{L_1(\mathbb{R}^n, \mathcal{B}_1)}}{\beta}.$$

Since compactly supported functions are dense in $L_1(\mathbb{R}^n, \mathcal{B}_1)$ by Lemma 1.3.3 we obtain continuity from $L_1(\mathbb{R}^n, B_1)$ to $L_{1,\infty}(\mathbb{R}^n, B_2)$ and the estimate (2.19).

An application of the Banach space valued *Marcinkiewicz interpolation theorem* yields continuity $T : L_p(\mathbb{R}^n, B_1) \to L_p(\mathbb{R}^n, B_2)$ for $1 < p \leq q$, hence the result follows for $q = \infty$. So we are left to consider the case of $q < \infty$. It is known that the adjoint operator T^* of T is a continuous operator from $L_{q'}(\mathbb{R}^n, B_2^*)$ to $L_{q'}(\mathbb{R}^n, B_1^*)$ with kernel $K^*(-x)$, where $K^* : \mathbb{R}^n \to \mathcal{L}(B_2^*, B_1^*)$, see Grafakos [41, Theorem 5.6.1] for details on these statements. As the kernels K and K^* have the same operator norm condition (2.18) is satisfied by K^*. Hence the previous arguments apply to T^* and we therefore obtain that T^* is bounded from $L_r(\mathbb{R}^n, B_2^*) \to L_r(\mathbb{R}^n, B_1^*)$ for any $1 < r \leq q'$. This implies boundedness for T from $L_p(\mathbb{R}^n, B_1) \to L_p(\mathbb{R}^n, B_2)$ for $q \leq p < \infty$ by the previous arguments and hence the results for all stated p. $\square$

With the preparations of Lemma 2.3.5 we generalize the ideas of Benedek, Calderón and Panzone [12] to show boundedness of convolution operators in the $L_{\vec{p}}$-setting. A similar theorem has been proven by Krée [58, Theorem 4], but as the proof is a bit sparse we will present a detailed proof.

Theorem 2.3.6. *Let $\mathcal{B}_1$ and $\mathcal{B}_2$ be Banach spaces and $K : \mathbb{R}^n \to \mathcal{L}(\mathcal{B}_1, \mathcal{B}_2)$ locally integrable. Furthermore we require that the operator*

$$TF(x) = \int_{\mathbb{R}^n} K(x - y)F(y)dy$$

is bounded from $L_{\vec{q}}(\mathbb{R}^n, \mathcal{B}_1)$ to $L_{\vec{q}}(\mathbb{R}^n, \mathcal{B}_2)$ for one $\vec{q} \in (1, \infty)^n$ with norm $B > 0$. Additionally, we assume that K fulfils an anisotropic Hörmander's condition

$$(2.21) \qquad \sup_{y \neq 0} \int_{|x|_{\vec{a}} \geq \eta|y|_{\vec{a}}} \|K(x - y) - K(x)\|_{\mathcal{L}(B_1, B_2)} \, dx \leq A$$

for some $\eta \in (1, \infty)$. Then the operator T extends to a continuous operator from $L_{\vec{p}}(\mathbb{R}^n, \mathcal{B}_1)$ to $L_{\vec{p}}(\mathbb{R}^n, \mathcal{B}_2)$ for all $\vec{p} \in (1, \infty)^n$ and satisfies the estimate

$$(2.22) \qquad \|TF\|_{L_{\vec{p}}(\mathbb{R}^n, \mathcal{B}_2)} \leq C(n, \vec{p}, A, B)\|F\|_{L_{\vec{p}}(\mathbb{R}^n, \mathcal{B}_1)}$$

for all $F \in L_{\vec{p}}(\mathbb{R}^n, \mathcal{B}_1)$ and some constant $C(n, \vec{p}, A, B) > 0$.

Proof: We argue by induction and for $n = 1$ the result follows from Lemma 2.3.5. Hence we can assume the result holds for $n - 1$ variables. Every function $F \in L_{\vec{p}}(\mathbb{R}^n, \mathcal{B}_1)$ can also be understood as a function in $L_{\vec{p}'}(\mathbb{R}^{n-1}, L_{p_n}(\mathbb{R}, \mathcal{B}_1))$ and to avoid confusion we denote by ΞF the function F understood as an element of $L_{\vec{p}'}(\mathbb{R}^{n-1}, L_{p_n}(\mathbb{R}, \mathcal{B}_1))$. We define

$$\gamma_l F := \begin{cases} F & \text{if } |x_n| \leq l, \\ 0 & \text{if } |x_n| > l, \end{cases} \quad \text{and} \quad K_l(x')h := \begin{cases} \int_{-l}^{l} K(x', x_n - y_n) h(y_n) \, dy_n & \text{if } |x_n| \leq l, \\ 0 & \text{if } |x_n| > l, \end{cases}$$

for $h \in L_{p_n}(\mathbb{R}, \mathcal{B}_1)$ and $l \in \mathbb{N}$. We conclude $K_l \in \mathcal{L}(L_{p_n}(\mathbb{R}, \mathcal{B}_1), L_{p_n}(\mathbb{R}, \mathcal{B}_2))$ for almost all $x' \in \mathbb{R}^{n-1}$ from

$$\|K_l(x')h\|_{L_{p_n}(\mathbb{R}, \mathcal{B}_2)} \leq \Big(\int_{-l}^{l} \Big| \int_{-l}^{l} \|K(x', x_n - y_n)\|_{\mathcal{L}(\mathcal{B}_1, \mathcal{B}_2)} \|h(y_n)\|_{\mathcal{B}_1} \, dy_n \Big|^{p_n} \, dx_n \Big)^{\frac{1}{p_n}}$$

$$\leq \Big(\int_{-l}^{l} \Big| \int_{-2l}^{2l} \|K(x', z_n)\|_{\mathcal{L}(\mathcal{B}_1, \mathcal{B}_2)} \|h(x_n - z_n)\|_{\mathcal{B}_1} \, dz_n \Big|^{p_n} \, dx_n \Big)^{\frac{1}{p_n}}$$

$$\leq \|h\|_{L_{p_n}(\mathbb{R}, \mathcal{B}_1)} \int_{-2l}^{2l} \|K(x', z_n)\|_{\mathcal{L}(\mathcal{B}_1, \mathcal{B}_2)} \, dz_n.$$

The last estimate is an application of Minkowski's integral inequality, and the inequality implies that K_l is locally integrable as K was. Therefore, we conclude

$$\int_{|x'|_{\vec{a}'} \geq \eta |y'|_{\vec{a}'}} \|K_l(x' - y') - K_l(x')\|_{\mathcal{L}(L_{p_n}(\mathbb{R}, \mathcal{B}_1), L_{p_n}(\mathbb{R}, \mathcal{B}_2))} \, dx'$$

$$\leq \int_{|x'|_{\vec{a}'} \geq \eta |y'|_{\vec{a}'}} \int_{\mathbb{R}} \|K(x' - y', x_n) - K(x', x_n)\|_{\mathcal{L}(\mathcal{B}_1, \mathcal{B}_2)} \, dx_n \, dx'$$

$$\leq \int_{|(x', x_n)|_{\vec{a}} \geq \eta |(y', 0)|_{\vec{a}}} \|K((x', x_n) - (y', 0)) - K(x)\|_{\mathcal{L}(\mathcal{B}_1, \mathcal{B}_2)} \, dx \leq A$$

by inequality (2.21).

For functions $F \in L_\infty^0(\mathbb{R}^{n-1}, L_{p_n}(\mathbb{R}, \mathcal{B}_1))$ we define the operator

$$T_l F(x') := \int_{\mathbb{R}^{n-1}} K_l(x' - y') F(y') \, dy',$$

which is well-defined by the local integrability of K_l. For a function $H \in L_\infty^0(\mathbb{R}^n, \mathcal{B}_1)$ we derive the identity $T_l \Xi H = \Xi \gamma_l T \gamma_l H$ and hence the estimate

(2.23) $$\|T_l \Xi H\|_{L_{\vec{q}'}(\mathbb{R}^{n-1}, L_{q_n}(\mathbb{R}, \mathcal{B}_2))} = \|\gamma_l T \gamma_l H\|_{L_{\vec{q}}(\mathbb{R}^n, \mathcal{B}_2)} \leq \|T \gamma_l H\|_{L_{\vec{q}}(\mathbb{R}^n, \mathcal{B}_2)}$$

$$\leq B \|\Xi H\|_{L_{\vec{q}'}(\mathbb{R}^{n-1}, L_{q_n}(\mathbb{R}, \mathcal{B}_1))}.$$

By Lemma 1.3.3 b) the set $L_\infty^0(\mathbb{R}^n, \mathcal{B}_1)$ is dense in $L_{\vec{q}'}(\mathbb{R}^{n-1}, L_{q_n}(\mathbb{R}, \mathcal{B}_1))$ and hence for every l the operator T_l is bounded with norm estimated by B independent of l. The operators T_l satisfy inequality (2.21) uniformly in l and an application of the induction assumption yields

$$\|T_l G\|_{L_{\vec{r}}(\mathbb{R}^{n-1}, L_{q_n}(\mathbb{R}, \mathcal{B}_2))} \leq C(n - 1, \vec{r}, A, B) \|G\|_{L_{\vec{r}}(\mathbb{R}^{n-1}, L_{q_n}(\mathbb{R}, \mathcal{B}_1))}$$

for all $G \in L_{\vec{r}}(\mathbb{R}^{n-1}, L_{q_n}(\mathbb{R}, \mathcal{B}_1))$ and all $\vec{r} \in (1, \infty)^{n-1}$. This implies

$$\|\gamma_l T \gamma_l F\|_{L_{\vec{r},q_n}(\mathbb{R}^n, \mathcal{B}_2)} = \|T_l \Xi F\|_{L_{\vec{r}}(\mathbb{R}^{n-1}, L_{q_n}(\mathbb{R}, \mathcal{B}_2))}$$
$$\leq C(n-1, \vec{r}, A, B) \|\Xi F\|_{L_{\vec{r}}(\mathbb{R}^{n-1}, L_{q_n}(\mathbb{R}, \mathcal{B}_1))} = C(n-1, \vec{r}, A, B) \|F\|_{L_{\vec{r},q_n}(\mathbb{R}^n, \mathcal{B}_1)}$$

for any $F \in L_{\vec{r},q_n}(\mathbb{R}^n, \mathcal{B}_1)$ and every $\vec{r} \in (1, \infty)^{n-1}$.

We are left to show that $\gamma_l T \gamma_l F$ converges to TF for $F \in L^0_\infty(\mathbb{R}^n, \mathcal{B}_1)$ in $L_{\vec{r},q_n}(\mathbb{R}^n, \mathcal{B}_2)$. Note that $\gamma_l F = F$ for l large enough and an application of the monotone convergence theorem yields $\|TF\|_{L_{\vec{r},q_n}(\mathbb{R}^n, \mathcal{B}_2)} \lesssim \|F\|_{L_{\vec{r},q_n}(\mathbb{R}^n, \mathcal{B}_1)}$ for $r \in (1, \infty)^{n-1}$. Hence we can apply the theorem of dominated convergence to conclude that $\gamma_l T \gamma_l F$ converges to TF in $L_{\vec{r},q_n}(\mathbb{R}^n, \mathcal{B}_2)$ for any $F \in L^0_\infty(\mathbb{R}^n, \mathcal{B}_1)$, and since the operators are uniformly bounded, this implies convergence for all $F \in L_{\vec{r},q_n}(\mathbb{R}^n, \mathcal{B}_1)$.

To be able to modify q_n we pick $\vec{r}$ to be equal to q_n in every component and apply Lemma 2.3.5 to obtain continuity for all $p \in (1, \infty)$. For a given $\vec{p} \in (1, \infty)^n$ we fix p_n and use the $L_{p_n}(\mathbb{R}^n, \mathcal{B}_1)$-continuity of T together with the previous steps to deduce $L_{\vec{p}}(\mathbb{R}^n, \mathcal{B}_1)$ continuity. Since the last steps only require applications of Lemma 2.3.5 and the previously used arguments we derive the stated dependence of the constant. $\qquad \square$

Remark 2.3.7. The above approach to derive $L_{\vec{p}}$ estimates is quite standard, but fails in the case where the function K is singular, because the operators K_l do in general not have to exist. A way to consider singular integral operators is either the theory of Muckenhoupt weights, see for example Kurtz [59], or $\mathcal{R}$-boundedness, see for example Hytönen and Portal [47].

Motivated by the ideas of Krée [58] we state an alternative condition to the one of Hörmander from (2.18) for the function K to satisfy to apply the previous Lemmata.

Corollary 2.3.8. For $B_j := B_{\vec{a}}(0, 2^j)$ the condition from (2.18) or (2.21) can be replaced by the condition, that there exists an $m \in \mathbb{N}$ such that

$$(2.24) \qquad \sup_{j \in \mathbb{Z}} \int_{\mathbb{R}^n \setminus B_j} \|K(x-y) - K(x)\|_{\mathcal{L}(\mathcal{B}_1, \mathcal{B}_2)} \, \mathrm{d}x \leq A \quad \text{for all } y \in B_{j-m}.$$

Proof: For any $y \in \mathbb{R}^n \setminus \{0\}$ we find a unique $j \in \mathbb{Z}$ such that $y \in B_{j-m}$ but $y \notin B_{j-1-m}$. For $\eta = 2^{m+1}$ every x of the set $\{x \in \mathbb{R}^n \mid |x|_{\vec{a}} \geq \eta |y|_{\vec{a}}\}$ satisfies

$$|x|_{\vec{a}} \geq 2^{m+1} |y|_{\vec{a}} \geq 2^{m+1} 2^{j-1-m} = 2^j,$$

and therefore $x \in \mathbb{R}^n \setminus B_j$. Hence the integral from (2.24) is a majorant to the integral from (2.18) or (2.21). As the choice of η does not depend on y we obtain the result. $\qquad \square$

2.3.2 The Decomposition Result

With the preparations of the previous section we are able to prove an anisotropic Littlewood-Paley decomposition and we start by introducing an anisotropic decomposition of unity. As this decomposition will be of importance in Chapter 3 we will be thorough in its introduction and introduce more than strictly needed for this section.

Fix $\vec{a} \in (0, \infty)^n$ and $b \in (0, \infty)$ throughout this section and let $\tau \in C^\infty_0(\mathbb{R})$ be a positive function bounded by 1 such that

$$\tau(t) = \begin{cases} 1 & |t| \leq c, \\ 0 & |t| \geq d, \end{cases}$$

where $c < d$ and $c, d \in \mathbb{R}_+$. Next set $\tau_j : \mathbb{R}^{n+1} \to \mathbb{R}$ as $\tau_j(\xi) = \tau(2^{-j}|\xi|_{b,\vec{a}})$ for $j \in \mathbb{N}_0$ to define functions $\psi_j \in C_0^\infty(\mathbb{R}^{n+1})$ by

$$(2.25) \qquad \begin{aligned} \psi_j(\xi) &= \tau_j(\xi) - \tau_{j-1}(\xi) \quad \text{for } j \in \mathbb{N}, \\ \psi_0(\xi) &= \tau_0(\xi). \end{aligned}$$

The identity $1 = \sum_{j=0}^\infty \psi_j(\xi)$ holds for all $\xi \in \mathbb{R}^{n+1}$ and

$$\begin{aligned} \operatorname{supp} \psi_j &\subset \left\{ \xi \in \mathbb{R}^{n+1} \big| \, c2^{j-1} \leq |\xi|_{b,\vec{a}} \leq d2^j \right\} \quad j \in \mathbb{N}, \\ \operatorname{supp} \psi_0 &\subset \left\{ \xi \in \mathbb{R}^{n+1} \big| \, |\xi|_{b,\vec{a}} \leq d \right\}. \end{aligned}$$

As long as $2c > d$ we define $\gamma \in C_0^\infty(\mathbb{R})$ as a positive function bounded by 1 such that

$$\gamma(t) = \begin{cases} 1 & |t| \leq d, \\ 0 & |t| \geq 2c, \end{cases}$$

and in the same way as before $\gamma_j : \mathbb{R}^{n+1} \to \mathbb{R}$ by $\gamma_j(\xi) = \gamma(2^{-j}|\xi|_{b,\vec{a}})$ for $j \in \mathbb{Z}$ and $j \geq -1$. This enables the definition of functions $\Psi_j \in C_0^\infty(\mathbb{R}^{n+1})$ by

$$(2.26) \qquad \begin{aligned} \Psi_j(\xi) &= \gamma_j(\xi) - \gamma_{j-2}(\xi) \quad \text{for } j \in \mathbb{N}, \\ \Psi_0(\xi) &= \gamma_0(\xi), \end{aligned}$$

such that $\psi_j(\xi)\Psi_j(\xi) = \psi_j(\xi)$ holds for all $\xi \in \mathbb{R}^{n+1}$ and all $j \in \mathbb{N}_0$. This follows from the facts that γ_j equals 1 on $|\xi|_{b,\vec{a}} \leq d2^j$ and γ_{j-2} equals zero as long as $2^{j-1}c \leq |\xi|_{b,\vec{a}}$, hence we obtain the identity for $j \in \mathbb{N}$ and as the case of $j = 0$ is clear it holds for all $j \in \mathbb{N}_0$.

For the first decomposition result we need an additional set of functions $\phi_j \in C_0^\infty(\mathbb{R}^n)$, defined by

$$(2.27) \qquad \phi_j(\xi) = \tau(2^{-j}|\xi|_{\vec{a}}) - \tau(2^{1-j}|\xi|_{\vec{a}}) \quad \text{for } j \in \mathbb{Z};$$

here the assumption $2c > d$ does not need to be satisfied. Straightforward calculations show

$$(2.28) \qquad \operatorname{supp} \phi_j \subset \left\{ \xi \in \mathbb{R}^n \mid c2^{j-1} \leq |\xi|_{\vec{a}} \leq d2^j \right\} \quad \text{for } j \in \mathbb{Z},$$

and $\sum_{j \in \mathbb{Z}} \phi_j(\xi) = 1$ for all $\xi \in \mathbb{R}^n \setminus \{0\}$. The following proof extends the ideas of Lizorkin [65] to the anisotropic case, and by the structure of ϕ_j we avoid a small inaccuracy of the cited proof. It is known that by a more iterated approach, for example with the ideas of Grafakos [41, Section 6.1.3], the proof of Lizorkin can be adjusted.

Lemma 2.3.9. *Let $1 < \vec{p} < \infty$, and ϕ_j from (2.27) with arbitrary values of $c < d$. Then there exist constants $C(\vec{p}, \vec{a}) > 0$ such that*

$$(2.29) \qquad \left\| \left(\sum_{j \in \mathbb{Z}} |\mathscr{F}_{\mathbb{R}^n}^{-1} \phi_j \mathscr{F}_{\mathbb{R}^n} f|^2 \right)^{\frac{1}{2}} \right\|_{L_{\vec{p}}(\mathbb{R}^n)} \leq C(\vec{p}, \vec{a}) \|f\|_{L_{\vec{p}}(\mathbb{R}^n)}$$

for all $f \in L_{\vec{p}}(\mathbb{R}^n)$.

Proof: Lemma 2.3.9 holds for arbitrary dimension and the $n+1$-dimensional version of ϕ_j coincides with ψ_j for $j \in \mathbb{N}$ for arbitrary $(b, \vec{a}) \in (0, \infty)^{n+1}$. Hence the existence of $C_{\vec{p}}^2$ follows from Lemma 2.3.9 and the obvious estimate $\|\mathscr{F}^{-1}\psi_0\mathscr{F}f\|_{\vec{p}} \lesssim \|f\|_{\vec{p}}$ as $\mathscr{F}^{-1}\psi_0 \in \mathscr{S}(\mathbb{R}^{n+1})$ by a direct application of the triangle inequality.

For the opposite direction we take Ψ_j from (2.26) such that $\Psi_j\psi_j = \psi_j$ for all $j \in \mathbb{N}_0$. The estimate from the beginning of the proof holds true for Ψ_j as well, since we can apply Lemma 2.3.9 for $j \in \mathbb{N}$ with corresponding $\tilde{\phi}_j$ and the case of $j = 0$ is again obvious. We take $f \in \mathscr{S}(\mathbb{R}^{n+1})$ and define $f_N := \sum_{j=0}^N \mathscr{F}_{\mathbb{R}^{n+1}}^{-1}\psi_j\Psi_j\mathscr{F}_{\mathbb{R}^{n+1}}f$ for $N \in \mathbb{N}$. An application of Plancherel's theorem yields

$$\left\| \sum_{j=0}^N \mathscr{F}_{\mathbb{R}^{n+1}}^{-1}\psi_j\Psi_j\mathscr{F}_{\mathbb{R}^{n+1}}f - f \right\|_{\mathrm{L}_2(\mathbb{R}^{n+1})} = \left\| \sum_{j=0}^N \psi_j\mathscr{F}_{\mathbb{R}^{n+1}}f - \mathscr{F}_{\mathbb{R}^{n+1}}f \right\|_{\mathrm{L}_2(\mathbb{R}^{n+1})},$$

and hence by the theorem of dominated convergence we obtain convergence of f_N to f in $\mathrm{L}_2(\mathbb{R}^{n+1})$, since $\sum_{j=0}^N \psi_j$ converges to 1 by construction. This implies convergence in $\mathscr{S}'(\mathbb{R}^{n+1})$ and for $f, g \in \mathscr{S}(\mathbb{R}^{n+1})$ we obtain

(2.33)
$$\left| \int_{\mathbb{R}^{n+1}} f(x)\overline{g(x)}\,\mathrm{d}x \right| = |\langle f, \overline{g}\rangle| = \left| \sum_{j=0}^\infty \langle \mathscr{F}_{\mathbb{R}^{n+1}}^{-1}\Psi_j\psi_j\mathscr{F}_{\mathbb{R}^{n+1}}f, \overline{g}\rangle \right| = \left| \sum_{j=0}^\infty \langle \psi_j\mathscr{F}_{\mathbb{R}^{n+1}}f, \overline{\Psi_j\mathscr{F}_{\mathbb{R}^{n+1}}g}\rangle \right|$$

$$= \left| \sum_{j=0}^\infty \langle \mathscr{F}_{\mathbb{R}^{n+1}}^{-1}\psi_j\mathscr{F}_{\mathbb{R}^{n+1}}f, \overline{\mathscr{F}_{\mathbb{R}^{n+1}}^{-1}\Psi_j\mathscr{F}_{\mathbb{R}^{n+1}}g}\rangle \right|$$

$$\leq \int_{\mathbb{R}^{n+1}} \sum_{j=0}^\infty |\mathscr{F}_{\mathbb{R}^{n+1}}^{-1}[\psi_j\mathscr{F}_{\mathbb{R}^{n+1}}f](x)\overline{\mathscr{F}_{\mathbb{R}^{n+1}}^{-1}[\Psi_j\mathscr{F}_{\mathbb{R}^{n+1}}g](x)}|\,\mathrm{d}x$$

$$\leq \int_{\mathbb{R}^{n+1}} \left\|\mathscr{F}_{\mathbb{R}^n}^{-1}[\psi_j\mathscr{F}_{\mathbb{R}^n}f](x)\right\|_{\ell_2} \left\|\mathscr{F}_{\mathbb{R}^{n+1}}^{-1}[\Psi_j\mathscr{F}_{\mathbb{R}^{n+1}}g](x)\right\|_{\ell_2}\,\mathrm{d}x$$

$$\leq \left\|\left\|\mathscr{F}_{\mathbb{R}^{n+1}}^{-1}[\psi_j\mathscr{F}_{\mathbb{R}^{n+1}}f]\right\|_{\ell_2}\right\|_{\mathrm{L}_{\vec{p}}(\mathbb{R}^{n+1})}\left\|\left\|\mathscr{F}_{\mathbb{R}^{n+1}}^{-1}[\Psi_j\mathscr{F}_{\mathbb{R}^{n+1}}g]\right\|_{\ell_2}\right\|_{\mathrm{L}_{\vec{p}'}(\mathbb{R}^{n+1})}$$

$$\leq C(\vec{p}, (b, \vec{a}))\left\|\left\|\mathscr{F}_{\mathbb{R}^{n+1}}^{-1}[\psi_j\mathscr{F}_{\mathbb{R}^{n+1}}f]\right\|_{\ell_2}\right\|_{\mathrm{L}_{\vec{p}}(\mathbb{R}^{n+1})}\|g\|_{\mathrm{L}_{\vec{p}'}(\mathbb{R}^{n+1})}.$$

This implies the opposite direction by Lemma 1.3.2. $\qquad\square$

We close this chapter by proving the Littlewood-Paley decomposition on G_n, but we are going to show a bit more than in $\mathbb{R}^{n+1}$, as we will show that any distribution $u \in \mathscr{S}'(G_n)$ with finite norm of the decomposition is in fact a regular distribution given by an element of $\mathrm{L}_{q,\vec{p}}(G_n)$.

Corollary 2.3.11. Let $1 < q, \vec{p} < \infty$, and $\varphi_j := \psi_j|_{\widehat{G}_n}$ with ψ_j as in (2.25) with $2c > d$. Then there exist constants $C_{q,\vec{p}}^1, C_{q,\vec{p}}^2$ such that

(2.34)
$$C_{q,\vec{p}}^1\|f\|_{\mathrm{L}_{q,\vec{p}}(G_n)} \leq \left\|\left(\sum_{j\in\mathbb{N}_0} |\mathscr{F}^{-1}\varphi_j\mathscr{F}f|^2\right)^{\frac{1}{2}}\right\|_{\mathrm{L}_{q,\vec{p}}(G_n)} \leq C_{q,\vec{p}}^2\|f\|_{\mathrm{L}_{q,\vec{p}}(G_n)}$$

for all $f \in \mathrm{L}_{q,\vec{p}}(G_n)$. Furthermore if the expression $\left\|(\mathscr{F}^{-1}\varphi_j\mathscr{F}u)_{j\in\mathbb{N}_0}\right\|_{\mathrm{L}_{q,\vec{p}}(G_n,\ell_2)}$ is finite for any $u \in \mathscr{S}'(G_n)$, then $u \in \mathrm{L}_{q,\vec{p}}(G_n)$ and u satisfies the estimate from (2.34).

Proof: The existence of $C^2_{q,\vec{p}}$ follows from Theorem 2.3.10 and Theorem 2.2.5 with $\mathcal{B}_2 = \ell_2$ and $\mathcal{B}_1 = \mathbb{C}$.

As the expression is finite for any $f \in \mathrm{L}_{q,\vec{p}}(G_n)$ by the existence of $C^2_{q,\vec{p}}$ we can restrict the considerations to the second statement. For $u \in \mathscr{S}'(G_n)$ the identity $u = \sum_{j=0}^{\infty} \mathscr{F}^{-1} \varphi_j \Phi_j \mathscr{F} u$ holds for $\Phi_j := \Psi_j\big|_{\widehat{G}_n}$ as $\sum_{j=0}^{\infty} \varphi_j \Phi_j = \sum_{j=0}^{\infty} \varphi_j = 1$. By repeating the steps of (2.33) we conclude the estimate

$$|\langle u, \overline{\varphi} \rangle| \lesssim \left\| \left\| (\mathscr{F}^{-1} \varphi_j \mathscr{F} u)_{j \in \mathbb{N}_0} \right\|_{\ell_2} \right\|_{\mathrm{L}_{q,\vec{p}}(G_n)} \|\varphi\|_{\mathrm{L}_{q',\vec{p'}}(G_n)},$$

for every $\varphi \in \mathscr{S}(G_n)$. Lemma 1.3.2 yields the result since u is a continuous functional on $\mathrm{L}_{q',\vec{p'}}(G_n)$ and hence an element of $\mathrm{L}_{q,\vec{p}}(G_n)$ which satisfies the estimate. $\qquad\square$

CHAPTER 3

Time-Periodic Anisotropic Function Spaces

An important tool for the analysis of boundary value problems are estimates of a trace operator. In the parabolic setting, the work of Weidemaier [81–83] and additionally the paper of Denk, Hieber and Prüss [27] determined the trace space of $W_{q,p}^{1,2}((0,T) \times \Omega)$ for some time $T > 0$ and a regular domain $\Omega \subset \mathbb{R}^n$. They expressed this trace space as an intersection of Triebel-Lizorkin spaces with values in a Lebesgue or Sobolev-Slobodeckij space. Later the study of this problem was generalized to the trace problem of arbitrary Triebel-Lizorkin spaces, see Johnsen and Sickel [56] and Johnsen, Munch Hansen and Sickel [53, 54] for details. They were not the first to consider this type of problem, since Berkolaïko [14–17] derived results previously. A good overview of these results can be found in Denk and Kaip [28, Section 3.2].

The goal of this chapter is to extend the results of [56] to time-periodic functions. Before we continue it is important to note that the previous results cannot directly be applied to the time-periodic setting because given a function u in the trace space the cited theory constructs an extension Eu such that its trace is u, but the periodicity of Eu cannot be assured. This operator E is called a right inverse to the trace operator Tr because we have $\mathrm{Tr} \circ E = \mathrm{Id}$. For the considerations in Chapter 4 we need an answer to the question "If u is time-periodic, is it possible to construct a time-periodic Eu?". Since the cited work cannot give an answer to this question more considerations are necessary. Additionally, we need to be able to construct extensions that preserve pure periodicity, i.e., if the boundary value is an element of the image of the projection $\mathcal{P}_\perp$, the constructed right inverse needs to be as well. All of this led to the considerations of Celik and Kyed [20, 21], who studied this problem in the case $W_p^{1,2}(\mathbb{T} \times \Omega)$. Observe that their approach cannot be extended to the case of $W_{q,p}^{1,2}$ with $p \neq q$ which is why we have to proceed in a different way in the following.

In Section 3.1 we introduce time-periodic anisotropic Besov and Triebel-Lizorkin spaces on $G_n = \mathbb{T} \times \mathbb{R}^n$ using the anisotropic distance function from Section 1.4, which was formalized by Yamazaki [85,86]. To our knowledge such spaces have not been considered on a combination of $\mathbb{T}$ and $\mathbb{R}$ before, for early considerations on either $\mathbb{R}^n$ or $\mathbb{T}^n$ see Triebel [80, Chapter 10] or Schmeisser and Triebel [74] for example. To keep this thesis self-contained, we will refrain from omitting similar proofs. A reader familiar with this type of spaces can easily skim over this section.

Section 3.2 concerns the question of whether the product of two functions from Besov or Triebel-Lizorkin spaces is an element of any space of this type. While this question is easily answered for $L_{\vec{p}}$ functions by Hölder's inequality (1.4), it is fairly technical in this general setting. We refer to

Remark 3.1.6. We will not study these spaces in any detail, but the needed results are provided in the appendix, for example a justification of why these are called homogeneous can be found in Lemma 5.1.6.

We briefly remark in which way these spaces defined above can be viewed as extension to the ones defined on $\mathbb{R}^n$ in Johnsen and Sickel [56]. By setting $k = 0$ in the definition of φ_j for all $j \in \mathbb{N}_0$, we get an anisotropic decomposition corresponding to $\vec{a}$ in $\mathbb{R}^n$. A distribution $u \in \mathscr{S}'(\mathbb{R}^n)$ can be extended in natural way to a distribution $v \in \mathscr{S}'(G_n)$ by $v = 1 \otimes u$. The following corollary shows that the anisotropic norms on $\mathbb{R}^n$ and G_n coincide for u and v.

Corollary 3.1.7. For $u \in \mathscr{S}'(\mathbb{R}^n)$ the identities

$$\left\| 1 \otimes u \right\|_{\mathrm{F}^{s,(b,\vec{a})}_{(q,\vec{p}),r}(G_n)} = \left\| u \right\|_{\mathrm{F}^{s,\vec{a}}_{\vec{p},r}(\mathbb{R}^n)} := \left\| \left(\sum_{j=0}^{\infty} 2^{jsr} \left| \mathscr{F}^{-1}_{\mathbb{R}^n}[\varphi_j(0,\cdot)\mathscr{F}_{\mathbb{R}^n} u](\cdot) \right|^r \right)^{\frac{1}{r}} \right\|_{\mathrm{L}_{\vec{p}}(\mathbb{R}^n)}$$

$$\left\| 1 \otimes u \right\|_{\mathrm{B}^{s,(b,\vec{a})}_{(q,\vec{p}),r}(G_n)} = \left\| u \right\|_{\mathrm{B}^{s,\vec{a}}_{\vec{p},r}(\mathbb{R}^n)} := \left(\sum_{j=0}^{\infty} 2^{jsr} \left\| \mathscr{F}^{-1}_{\mathbb{R}^n}[\varphi_j(0,\cdot)\mathscr{F}_{\mathbb{R}^n} u] \right\|^r_{\mathrm{L}_{\vec{p}}(\mathbb{R}^n)} \right)^{\frac{1}{r}}$$

hold for all $1 \leq r \leq \infty$, $s \in \mathbb{R}$, $1 \leq \vec{p}, q < \infty$ for Triebel-Lizorkin spaces, and $1 \leq \vec{p}, q \leq \infty$ for Besov spaces.

Proof: As $v = 1 \otimes u$ we get $\mathscr{F}v = \delta_0 \otimes \mathscr{F}_{\mathbb{R}^n} u$. Therefore, for $\varphi \in \mathscr{S}(G_n)$ it holds

$$\langle \mathscr{F}^{-1}[\varphi_j \mathscr{F} v], \varphi \rangle = \langle \delta_0 \otimes \mathscr{F}_{\mathbb{R}^n} u, \varphi_j \mathscr{F}^{-1}_{\widehat{G_n}} \varphi \rangle = \langle \mathscr{F}_{\mathbb{R}^n} u, \varphi_j(0,\cdot)[\mathscr{F}^{-1}_{\widehat{G_n}} \varphi](0,\cdot) \rangle$$

$$= \langle \delta_0 \otimes [\varphi_j(0,\cdot)\mathscr{F}_{\mathbb{R}^n} u], \mathscr{F}^{-1}_{\widehat{G_n}} \varphi \rangle = \langle \mathscr{F}^{-1}(\delta_0 \otimes [\varphi_j(0,\cdot)\mathscr{F}_{\mathbb{R}^n} u]), \varphi \rangle$$

$$= \langle \mathscr{F}^{-1}_{\mathbb{R}^n}[\varphi_j(0,\cdot)\mathscr{F}_{\mathbb{R}^n} u], \varphi \rangle.$$

This implies $\mathscr{F}^{-1}[\varphi_j \mathscr{F} v] = \mathscr{F}^{-1}_{\mathbb{R}^n}[\varphi_j(0,\cdot)\mathscr{F}_{\mathbb{R}^n} u]$ and therefore each element of the sum does not depend on t and hence the identities follow as the additional L_q-norm with respect to the torus equals 1. $\qquad\square$

We start the examination of anisotropic time-periodic function spaces on G_n with simple embeddings and completeness results. Although these results may be simple in some cases, they are important since we are going to frequently use them throughout this thesis. In $\mathbb{R}^n$ the results are well-known for isotropic Besov and Triebel-Lizorkin spaces and can be found in any textbook covering this topic.

Theorem 3.1.8. Let $s \in \mathbb{R}$, $1 \leq r \leq \infty$, and $1 \leq \vec{p}, q < \infty$ for Triebel-Lizorkin spaces or $1 \leq \vec{p}, q \leq \infty$ for Besov spaces.

a) For $\varepsilon > 0$ and $1 < r_1 < r_2 < \infty$ we get

$$\mathrm{F}^{s+\varepsilon,(b,\vec{a})}_{(q,\vec{p}),\infty}(G_n) \hookrightarrow \mathrm{F}^{s,(b,\vec{a})}_{(q,\vec{p}),1}(G_n) \hookrightarrow \mathrm{F}^{s,(b,\vec{a})}_{(q,\vec{p}),r_1}(G_n)$$

$$\hookrightarrow \mathrm{F}^{s,(b,\vec{a})}_{(q,\vec{p}),r_2}(G_n) \hookrightarrow \mathrm{F}^{s,(b,\vec{a})}_{(q,\vec{p}),\infty}(G_n) \hookrightarrow \mathrm{F}^{s-\varepsilon,(b,\vec{a})}_{(q,\vec{p}),1}(G_n).$$

The same result holds in the case of Besov spaces.

b) It holds

$$F^{s,(b,\vec{a})}_{(q,\vec{p}),r}(G_n) \hookrightarrow B^{s,(b,\vec{a})}_{(q,\vec{p}),r}(G_n)$$

for $\max\{q, p_1, p_2, \ldots, p_n\} \le r$.

c) It holds

$$B^{s,(b,\vec{a})}_{(q,\vec{p}),r}(G_n) \hookrightarrow F^{s,(b,\vec{a})}_{(q,\vec{p}),r}(G_n)$$

for $r \le \min\{q, p_1, p_2, \ldots, p_n\}$.

d) We get

$$\mathscr{S}(G_n) \hookrightarrow B^{s,(b,\vec{a})}_{(q,\vec{p}),r}(G_n) \hookrightarrow \mathscr{S}'(G_n).$$

The same result holds in the case of Triebel-Lizorkin spaces.

e) $B^{s,(b,\vec{a})}_{(q,\vec{p}),r}(G_n)$ and $F^{s,(b,\vec{a})}_{(q,\vec{p}),r}(G_n)$ are Banach spaces.

Proof:

a) The monotonicity of ℓ_q-spaces, i.e., $\|\cdot\|_{\ell_{r_2}} \le \|\cdot\|_{\ell_{r_1}}$ for all $1 \le r_1 \le r_2 \le \infty$, yields the embeddings in the cases where s does not change. For the remaining embeddings note that

$$\|f\|_{F^{s,(b,\vec{a})}_{(q,\vec{p}),1}(G_n)} = \left\| \|2^{js} \mathscr{F}^{-1}[\varphi_j \mathscr{F} f]\|_{\ell_1} \right\|_{L_{q,\vec{p}}(G_n)} \le \left\| \|2^{j(s+\varepsilon)} \mathscr{F}^{-1}[\varphi_j \mathscr{F} f]\|_{\ell_\infty} \|2^{-j\varepsilon}\|_{\ell_1} \right\|_{L_{q,\vec{p}}(G_n)}$$

$$= c\|f\|_{F^{s+\varepsilon,(b,\vec{a})}_{(q,\vec{p}),\infty}(G_n)}.$$

By the same argument, the result follows for Besov spaces.

b) The proof relies on applying Minkowski's integral inequality. If $r < \infty$, it holds

(3.3)
$$\left(\sum_{j=0}^{\infty} \left(\int |f_j(y)|^s \mathrm{d}y \right)^{\frac{r}{s}} \right)^{\frac{1}{r}} \le \left(\int \left(\sum_{j=0}^{\infty} |f_j(y)|^r \right)^{\frac{s}{r}} \mathrm{d}y \right)^{\frac{1}{s}}$$

for integration over $\mathbb{R}$ if $s = p_i \le r$ or for integration over $\mathbb{T}$ if $s = q \le r$. Hence by applying the inequality (3.3) $(n+1)$-times we obtain

$$F^{s,(b,\vec{a})}_{(q,\vec{p}),r}(G_n) \hookrightarrow B^{s,(b,\vec{a})}_{(q,\vec{p}),r}(G_n)$$

for $\max\{q, p_1, p_2, \ldots, p_n\} \le r$. Since $\sup_{j \in \mathbb{N}_0} \|f_j\|_{L_s} \le \|\sup_{j \in \mathbb{N}_0} |f_j|\|_{L_s}$ holds for any $1 \le s \le \infty$ we obtain the result for $r = \infty$.

c) This follows from the same steps as in b) but for $r \le s$ the estimate in inequality (3.3) is in the opposite way.

d) We first consider $B^{s,(b,\vec{a})}_{(q,\vec{p}),\infty}(G_n)$. For $f \in \mathscr{S}(G_n)$ it holds

$$
\begin{aligned}
\|f\|_{B^{s,(b,\vec{a})}_{(q,\vec{p}),\infty}(G_n)} &= \sup_{j\in\mathbb{N}_0} 2^{js}\|\mathscr{F}^{-1}[\varphi_j\mathscr{F}f]\|_{L_{q,\vec{p}}(G_n)} \\
&\leq \sup_{j\in\mathbb{N}_0} 2^{js}\|(1+|x|^2)^n\mathscr{F}^{-1}[\varphi_j\mathscr{F}f]\|_{L_\infty(G_n)}\|(1+|x|^2)^{-n}\|_{L_{q,\vec{p}}(G_n)} \\
&\lesssim \sup_{j\in\mathbb{N}_0} 2^{js}\|\mathscr{F}^{-1}[(1-c_n\Delta)^n(\varphi_j\mathscr{F}f)]\|_{L_\infty(G_n)} \\
&\lesssim \sum_{|\alpha|\leq 2n} \sup_{j\in\mathbb{N}_0} 2^{js}\|\mathscr{F}^{-1}[D^\alpha(\varphi_j\mathscr{F}f)]\|_{L_\infty(G_n)} \\
&\lesssim \sum_{|\alpha|\leq 2n} \sup_{j\in\mathbb{N}_0} 2^{js}\left\|\sum_{\beta\leq\alpha}\binom{\alpha}{\beta}D^\beta\varphi_j D^{\alpha-\beta}\mathscr{F}f\right\|_{L_1(\widehat{G}_n)}
\end{aligned}
$$

(3.4)

By (3.2) the functions φ_j and their derivatives are uniformly bounded with respect to j. Furthermore, we get $2^{js} \leq 2^s|(k,\xi)|^{\max\{0,s\}}_{b,\vec{a}}$ on the support of φ_j for any $j \in \mathbb{N}$. For $j = 0$ we have $2^{js} = 1$. There exist $m, l \in \mathbb{N}$ such that $s \leq m$ and $\tau^{-1} \leq l$ with $\tau := \min\{1, b, a_1, \ldots, a_n\}$. By rearranging the derivatives of $\mathscr{F}f$, this yields

$$
\begin{aligned}
\|f\|_{B^{s,(b,\vec{a})}_{(q,\vec{p}),\infty}(G_n)} &\lesssim \sum_{|\gamma|\leq 2n}\|D^\gamma(\mathscr{F}f)(1+|(k,\xi)|^{\max\{0,s\}}_{b,\vec{a}})\|_{L_1(\widehat{G}_n)} \\
&\lesssim \sum_{|\gamma|\leq 2n}\sup_{(k,\xi)\in\widehat{G}_n}|(1+|\xi|^2)^{n+ml}(1+k^2)^{ml+1}(D^\gamma\mathscr{F}f)(k,\xi)| \\
&\lesssim \sum_{\substack{|\alpha|\leq 2n+2ml \\ |\beta|\leq 2ml+2 \\ |\gamma|\leq 2n}}\sup_{(k,\xi)\in\widehat{G}_n}\left|\mathscr{F}\left[x^\gamma\partial_t^\beta\partial_x^\alpha f\right](k,\xi)\right| \\
&\lesssim \sum_{\substack{|\alpha|\leq 2n+2ml \\ \beta\leq 2ml+2 \\ |\gamma|\leq 4n}}\rho_{\alpha,\beta,\gamma}(f).
\end{aligned}
$$

(3.5)

Here we used $\frac{1+|(k,\xi)|^{\max\{0,s\}}_{b,\vec{a}}}{(1+k^2)^{ml}(1+|\xi|^2)^{n+ml}} \in L_1(\widehat{G}_n)$ since it holds

$$
1 + |(k,\xi)|^{\max\{0,s\}}_{b,\vec{a}} \lesssim 1 + |k|^{\max\{0,s\}l} + \sum_{j=1}^n |\xi_j|^{\max\{0,s\}l}
$$

and therefore the addition of n or 1 in the exponents of terms in the denominator ensures convergence. Furthermore we made use of properties of the Fourier transform with respect to derivatives and the estimate

$$
\sup_{(k,\xi)\in\widehat{G}_n}\left|\mathscr{F}\left[x^\gamma\partial_t^\beta\partial_x^\alpha f\right](k,\xi)\right| \leq \sup_{(t,x)\in G_n}|(1+|x|^2)^n x^\gamma\partial_t^\beta\partial_x^\alpha f(t,x)| \sup_{(k,\xi)\in\widehat{G}_n}\mathscr{F}[(1+|x|^2)^{-n}].
$$

Since the semi-norms $\rho_{\alpha,\beta,\gamma}$ generate the topology of $\mathscr{S}(G_n)$, the continuous embedding of $B^{s,(b,\vec{a})}_{(q,\vec{p}),\infty}(G_n)$ into $\mathscr{S}'(G_n)$ follows. The case of a general Besov or Triebel-Lizorkin space follows from a) and c).

By the same argument this time applying a) and b), we can restrict ourselves to the case $u \in B^{s,(b,\vec{a})}_{(q,\vec{p}),\infty}(G_n)$ for the remaining embedding. Because the support of φ_j is contained in $|(k,\xi)|_{b,\vec{a}} \leq 2^{j+1}$, Proposition 1.4.2 d) yields $|k| \leq 2^{jb}2^b$ and $|\xi_i| \leq 2^{ja_i}2^{a_i}$ for (k,ξ) in supp φ_j. Therefore, applying Lemma 2.1.1 with $K = 2^{jb}2^b$ and $R_i = 2^{ja_i}2^{a_i}$ for $1 \leq i \leq n$, we obtain

$$\|\mathscr{F}^{-1}[\varphi_j\mathscr{F}u]\|_{L_\infty(G_n)} \lesssim 2^{j(\frac{\vec{a}}{\vec{p}}+\frac{b}{q})}\|\mathscr{F}^{-1}\varphi_j\mathscr{F}u\|_{L_{(q,\vec{p})}(G_n)},$$

where $\frac{\vec{a}}{\vec{p}} = \sum_{j=1}^n \frac{a_j}{p_j}$. Let $\varphi \in \mathscr{S}(G_n)$ and Φ_j from Definition 3.1.2. It holds

$$|\langle u, \varphi \rangle| = \left|\left\langle \mathscr{F}u, \sum_{j=0}^\infty \varphi_j\Phi_j\mathscr{F}^{-1}_{\widehat{G}_n}\varphi \right\rangle\right| \leq \sum_{j=0}^\infty \left|\langle \mathscr{F}^{-1}\varphi_j\mathscr{F}u, \mathscr{F}_{\widehat{G}_n}[\Phi_j\mathscr{F}^{-1}_{\widehat{G}_n}\varphi]\rangle\right|$$

$$\leq \sum_{j=0}^\infty \|\mathscr{F}^{-1}\varphi_j\mathscr{F}u\|_{L_\infty(G_n)}\|\mathscr{F}_{\widehat{G}_n}[\Phi_j\mathscr{F}^{-1}_{\widehat{G}_n}\varphi]\|_{L_1(G_n)}$$

$$\lesssim \sum_{j=0}^\infty 2^{j(\frac{\vec{a}}{\vec{p}}+\frac{b}{q})}\|\mathscr{F}^{-1}\varphi_j\mathscr{F}u\|_{L_{(q,\vec{p})}(G_n)}\|\mathscr{F}_{\widehat{G}_n}[\Phi_j\mathscr{F}^{-1}_{\widehat{G}_n}\varphi]\|_{L_1(G_n)}$$

$$\lesssim \|u\|_{B^{s,(b,\vec{a})}_{(q,\vec{p}),\infty}} \sum_{j=0}^\infty 2^{j(-s+\frac{\vec{a}}{\vec{p}}+\frac{b}{q})}\|\mathscr{F}^{-1}\Phi_j\mathscr{F}\varphi\|_{L_1(G_n)}.$$

The interchange of $\mathscr{F}^{-1}$ and $\mathscr{F}^{-1}$ in the last estimate is possible due to $\Phi_j(-k,-\xi) = \Phi_j(k,\xi)$. The sum is the Besov norm of φ with Φ_j instead of φ_j. Since Φ_j satisfies estimate (3.2) and its support allows for the estimate $2^{js} \lesssim |(k,\xi)|_{b,\vec{a}}^{\max\{0,s\}}$, we can apply the previously shown estimates (3.4) and (3.5). Hence we find M such that

$$|\langle u, \varphi \rangle| \lesssim \|u\|_{B^{s,(b,\vec{a})}_{(q,\vec{p}),\infty}(G_n)} \sum_{|\alpha|+|\beta|+|\gamma|\leq M} \rho_{\alpha,\beta,\gamma}(\varphi).$$

This shows the continuous embedding into $\mathscr{S}'(G_n)$.

e) By definition it is clear that $B^{s,(b,\vec{a})}_{(q,\vec{p}),\infty}(G_n)$ and $F^{s,(b,\vec{a})}_{(q,\vec{p}),\infty}(G_n)$ are normed spaces. So it remains to prove completeness. Let $(u_l)_{l\in\mathbb{N}} \subset F^{s,(b,\vec{a})}_{(q,\vec{p}),r}(G_n)$ be a Cauchy sequence. By part d), $(u_l)_{l\in\mathbb{N}}$ is a Cauchy sequence in $\mathscr{S}'(G_n)$ as well and hence convergent to an element $u \in \mathscr{S}'(G_n)$ because $\mathscr{S}'(G_n)$ is sequentially complete. Due to φ_j being an element of $\mathscr{S}(\widehat{G}_n)$, the convolution of u with $\mathscr{F}^{-1}\varphi_j$ is defined by (1.2). We have

(3.6)
$$\mathscr{F}^{-1}[\varphi_j\mathscr{F}(u)](t,x) = c_n\langle u, \mathscr{F}^{-1}\varphi_j((t,x)-(\cdot,\cdot))\rangle = c_n \lim_{l\to\infty}\langle u_l, \mathscr{F}^{-1}\varphi_j((t,x)-(\cdot,\cdot))\rangle$$
$$= \lim_{l\to\infty}\mathscr{F}^{-1}[\varphi_j\mathscr{F}(u_l)](t,x).$$

Hence we obtain pointwise convergence of $\mathscr{F}^{-1}[\varphi_j\mathscr{F}u_l]$ for each $j \in \mathbb{N}_0$. We first consider the case $r < \infty$ and apply Fatou's Lemma $(n+2)$ times, which yields

$$\|u\|_{F^{s,(b,\vec{a})}_{(q,\vec{p}),r}(G_n)} \leq \liminf_{l\to\infty}\|u_l\|_{F^{s,(b,\vec{a})}_{(q,\vec{p}),r}(G_n)} < \infty.$$

This is possible as measurability at each step is guaranteed by the Theorem of Tonelli. To show convergence in $\mathrm{F}^{s,(b,\vec{a})}_{(q,\vec{p}),r}(G_n)$, we take arbitrary $\varepsilon > 0$ and find $N \in \mathbb{N}$ with the property that $\|u_m - u_i\|_{\mathrm{F}^{s,(b,\vec{a})}_{(q,\vec{p}),r}(G_n)} < \varepsilon$ for all $m, i \geq N$. Repeating the applications of Fatou's Lemma yields

$$\|u - u_i\|_{\mathrm{F}^{s,(b,\vec{a})}_{(q,\vec{p}),r}(G_n)} \leq \liminf_{m \to \infty} \|u_m - u_i\|_{\mathrm{F}^{s,(b,\vec{a})}_{(q,\vec{p}),r}(G_n)} < \varepsilon$$

and hence completeness for $r < \infty$. The case $r = \infty$ follows similarly, but instead of using Fatou to change sum and limit, we use the estimate

$$\sup_{j \in \mathbb{N}} |\mathscr{F}^{-1}[\varphi_j \mathscr{F}(u)](t,x)| \leq \liminf_{l \to \infty} \sup_{j \in \mathbb{N}} |\mathscr{F}^{-1}[\varphi_j \mathscr{F}(u_l)](t,x)|.$$

This estimate can be easily seen from (3.6), since the limit exists and we therefore derive

$$|\mathscr{F}^{-1}[\varphi_j \mathscr{F}(u)](t,x)| \leq \liminf_{l \to \infty} \sup_{j \in \mathbb{N}} |\mathscr{F}^{-1}[\varphi_j \mathscr{F}(u_l)](t,x)|.$$

This allows us to use the same steps to prove completeness for $r = \infty$. The same argument can be repeated to show completeness of the Besov spaces. $\qquad\square$

3.1.1 Independence of the Decomposition

A natural question concerning these spaces is whether the previous definitions depend on the choice of decomposition functions φ_j. Before we can answer this, we prepare some results. We follow the ideas of Marschall [67] and Johnsen and Sickel [56].

Lemma 3.1.9. *Let* $1 < \vec{p} < \infty$ *and* $1 < r \leq \infty$. *For* $\psi \in \mathscr{S}(\mathbb{R}^n)$ *and* $j \in \mathbb{N}$ *we set* $\psi_j = \psi(2^{-j\vec{a}} \cdot)$ *for* $j \in \mathbb{N}$. *Then there exists a constant* $c > 0$ *such that*

$$(3.7) \qquad \left\| \left\{ (\mathscr{F}^{-1}_{\mathbb{R}^n}[\psi_j \mathscr{F}_{\mathbb{R}^n} u_j]) \right\}_{j \in \mathbb{N}} \right\|_{\mathrm{L}_{\vec{p}}(\mathbb{R}^n, \ell_r)} \leq c \|\{u_j\}_{j \in \mathbb{N}}\|_{\mathrm{L}_{\vec{p}}(\mathbb{R}^n, \ell_r)}$$

for all elements $(u_j)_{j \in \mathbb{N}} \in \mathrm{L}_{\vec{p}}(\mathbb{R}^n, \ell_r)$.

Proof: As u_j is a regular distribution and $\psi_j \in \mathscr{S}(\mathbb{R}^n)$ for each $j \in \mathbb{N}$ we obtain the estimate

$$|\mathscr{F}^{-1}_{\mathbb{R}^n}[\psi_j \mathscr{F}_{\mathbb{R}^n} u_j](x)| \leq c_n \int_{\mathbb{R}^n} |[\mathscr{F}^{-1}_{\mathbb{R}^n} \psi_j](x-y) u_j(y)| \, \mathrm{d}y$$

$$= \sum_{k \in \mathbb{Z}} \int_{\mathbb{R}^n} |\phi_k(x-y)[\mathscr{F}^{-1}_{\mathbb{R}^n} \psi_j](x-y) u_j(y)| \, \mathrm{d}y,$$

where ϕ_k are the functions from equation (2.27). By the definition of ϕ_k we derive

$$\operatorname{supp} \phi_k \subset \{\xi \in \mathbb{R}^n \mid |\xi_i| \leq d^{a_i} 2^{k a_i} \text{ for all } 1 \leq i \leq n\} = Q_{(d2^k)^{\vec{a}}}(0).$$

Hence

$$(3.8) \qquad \begin{aligned} |\mathscr{F}^{-1}_{\mathbb{R}^n}[\psi_j \mathscr{F}_{\mathbb{R}^n} u_j](x)| &\lesssim \sum_{k \in \mathbb{Z}} \|\phi_k \mathscr{F}^{-1}_{\mathbb{R}^n} \psi_j\|_{\mathrm{L}_\infty(\mathbb{R}^n)} \int_{Q_{(d2^k)^{\vec{a}}}(x)} |u_j(y)| \mathrm{d}y \\ &\lesssim \sum_{k \in \mathbb{Z}} 2^{k|\vec{a}|} \|\mathscr{F}^{-1}_{\mathbb{R}^n}[\phi_k \mathscr{F}_{\mathbb{R}^n}[\psi_j(-\cdot)]]\|_{\mathrm{L}_1(\mathbb{R}^n)} M_S u_j(x) \\ &\lesssim \|\psi_j\|_{\widehat{\mathrm{B}}^{|\vec{a}|,\vec{a}}_{1,1}(\mathbb{R}^n)} M_S u_j(x), \end{aligned}$$

where M_S is the strong maximal operator defined in (1.23). Applying the scaling result of Lemma 5.1.6, the scaling factor vanishes in this case, we obtain

$$
(3.9) \qquad \left\| \left\{ \mathscr{F}^{-1}[\psi_j \mathscr{F} u_j](x) \right\}_{j \in \mathbb{N}} \right\|_{\ell_r} \lesssim \|\psi\|_{\widehat{\mathrm{B}}_{1,1}^{|\vec{a}|,\vec{a}}(\mathbb{R}^n)} \left\| \{ M_S u_j(x) \}_{j \in \mathbb{N}} \right\|_{\ell_r}.
$$

By Lemma 5.1.7 and Theorem 3.1.8 d) we see that $\|\psi\|_{\widehat{\mathrm{B}}_{1,1}^{|\vec{a}|,\vec{a}}(\mathbb{R}^n)}$ is finite. The result now follows from Lemma 1.7.3. $\qquad\square$

Remark 3.1.10. In the previous lemma it suffices that the required identity $\psi_j = \psi(2^{-j\vec{a}} \cdot)$ is only satisfied for $j \geq j_0$ for some $j_0 \in \mathbb{N}$, since the remaining finitely many can be estimated by Young's inequality (1.5) and then by the whole sum. We briefly show this statement, it holds

$$
\left\| \left\{ (\mathscr{F}_{\mathbb{R}^n}^{-1}[\psi_j \mathscr{F}_{\mathbb{R}^n} u_j] \right\}_{j \in \mathbb{N}} \right\|_{\mathrm{L}_{\vec{p}}(\mathbb{R}^n, \ell_r)} \lesssim \sum_{j=1}^{j_0} \|\mathscr{F}^{-1}\psi_j\|_{\mathrm{L}_1(\mathbb{R}^n)} \|u_j\|_{\mathrm{L}_{\vec{p}}(\mathbb{R}^n)} + \left\| \{u_j\}_{j \in \mathbb{N}} \right\|_{\mathrm{L}_{\vec{p}}(\mathbb{R}^n, \ell_r)}
$$

$$
\leq \sum_{j=1}^{j_0} C(j) \left\| \{u_k\}_{k \in \mathbb{N}} \right\|_{\mathrm{L}_{\vec{p}}(\mathbb{R}^n, \ell_r)} + \left\| \{u_j\}_{j \in \mathbb{N}} \right\|_{\mathrm{L}_{\vec{p}}(\mathbb{R}^n, \ell_r)}
$$

$$
\leq c(j_0) \left\| \{u_j\}_{j \in \mathbb{N}} \right\|_{\mathrm{L}_{\vec{p}}(\mathbb{R}^n, \ell_r)}.
$$

This implies that the inequality from (3.7) holds for φ_j and Φ_j.

We need additional convergence results in $\mathscr{S}'(G_n)$ and hence recall definitions made in Section 1.6. A sequence $(u_j)_{j \in \mathbb{N}_0}$ satisfies the *dyadic corona condition*, if there exist $A, B > 0$ such that

$$
(3.10) \qquad \operatorname{supp} \mathscr{F}_{G_n} u_j \subset \left\{ (k, \xi) \in \widehat{G}_n \,\middle|\, B 2^{j-1} \leq |(k, \xi)|_{b, \vec{a}} \leq A 2^j \right\},
$$

$$
\text{while } \operatorname{supp} \mathscr{F}_{G_n} u_0 \subset \left\{ (k, \xi) \in \widehat{G}_n \,\middle|\, |(k, \xi)|_{b, \vec{a}} \leq A \right\}.
$$

A sequence $(u_j)_{j \in \mathbb{N}_0}$ satisfies the *dyadic ball condition*, if there exists $A > 0$ such that for every $j \geq 0$

$$
(3.11) \qquad \operatorname{supp} \mathscr{F}_{G_n} u_j \subset \left\{ (k, \xi) \in \widehat{G}_n \,\middle|\, |(k, \xi)|_{b, \vec{a}} \leq A 2^j \right\}.
$$

As a preparation for Section 3.2 we extend the decomposition to the dyadic ball criterion in the case of $s \geq 0$.

Lemma 3.1.11. *Let $1 < \vec{p}, q < \infty$, and let $(u_j)_{j \in \mathbb{N}_0}$ be a sequence of distributions in $\mathscr{S}'(G)$ that fulfils*

$$
(3.12) \qquad \sum_{j=0}^{\infty} 2^{sj} \|u_j\|_{\mathrm{L}_{q,\vec{p}}(G_n)} < \infty.
$$

The following statements hold:

a) *If the sequence $(u_j)_{j \in \mathbb{N}_0}$ satisfies the condition from equation (3.11) for some $A > 0$ and $s \geq 0$, then the series $\sum_{j=0}^{\infty} u_j$ converges in $\mathrm{B}_{(q,\vec{p}),\infty}^{s,(b,\vec{a})}$ and in $\mathscr{S}'(G_n)$.*

b) *If the sequence $(u_j)_{j \in \mathbb{N}_0}$ satisfies the condition from equation (3.10) for some $A, B > 0$, then the series $\sum_{j=0}^{\infty} u_j$ converges in $\mathrm{B}_{(q,\vec{p}),\infty}^{s,(b,\vec{a})}(G_n)$ for each $s \in \mathbb{R}$ and in $\mathscr{S}'(G_n)$.*

Now repeating the steps from the proof of from Lemma 3.1.12, we derive the estimate

$$\Big\|\sum_{j=0}^{\infty} u_j\Big\|_{\mathrm{F}^{s,(b,\vec{a})}_{(q,\vec{p}),r}(G_n)} \lesssim \sum_{l=-h}^{h}\Big\|\Big(\sum_{k=0}^{\infty} 2^{ksr}|u_{l+k}|^r\Big)^{\frac{1}{r}}\Big\|_{\mathrm{L}_{q,\vec{p}}(G_n)}$$

$$\leq \sum_{l=-h}^{h} 2^{-ls}\Big\|\Big(\sum_{k=0}^{\infty} 2^{ksr}|u_k|^r\Big)^{\frac{1}{r}}\Big\|_{\mathrm{L}_{q,\vec{p}}(G_n)}$$

$$\lesssim \Big\|\Big(\sum_{k=0}^{\infty} 2^{ksr}|u_k|^r\Big)^{\frac{1}{r}}\Big\|_{\mathrm{L}_{q,\vec{p}}(G_n)}.$$

This proves the result for arbitrary $s \in \mathbb{R}$. $\qquad\square$

The previous results can easily be shown in the case of Besov spaces by reusing all the arguments above and interchanging the $\mathrm{L}_{(q,\vec{p})}$ and ℓ_r norms, where we apply Lemma 3.1.9 as before to sequences with single non-trivial entry, thus yielding the following lemma.

Lemma 3.1.14. *Let $1 < \vec{p}, q < \infty$, $1 < r \leq \infty$, and the sequence $(u_j)_{j\in\mathbb{N}_0}$ fulfil the condition*

$$\Big(\sum_{j=0}^{\infty} 2^{sjr}\|u_j\|^r_{\mathrm{L}_{q,\vec{p}}(G_n)}\Big)^{\frac{1}{r}} < \infty.$$

a) *If $s \in (0,\infty)$ and $(u_j)_{j\in\mathbb{N}_0}$ satisfies (3.11) for some $A > 0$, then the series $\sum_{j=0}^{\infty} u_j$ converges in $\mathscr{S}'(G_n)$ and in $\mathrm{B}^{s,(b,\vec{a})}_{(q,\vec{p}),r}(G_n)$, and there exists a constant $c > 0$ such that*

$$\Big\|\sum_{j=0}^{\infty} u_j\Big\|_{\mathrm{B}^{s,(b,\vec{a})}_{(q,\vec{p}),r}(G_n)} \leq c\Big(\sum_{j=0}^{\infty} 2^{sjr}\|u_j\|^r_{\mathrm{L}_{q,\vec{p}}(G_n)}\Big)^{\frac{1}{r}}.$$

b) *If $s \in \mathbb{R}$ and $(u_j)_{j\in\mathbb{N}_0}$ satisfies (3.10) for some $A, B > 0$, then the series $\sum_{j=0}^{\infty} u_j$ converges in $\mathscr{S}'(G_n)$ and in $\in \mathrm{B}^{s,(b,\vec{a})}_{(q,\vec{p}),r}(G_n)$ and there exists a constant $c > 0$ such that*

$$\Big\|\sum_{j=0}^{\infty} u_j\Big\|_{\mathrm{B}^{s,(b,\vec{a})}_{(q,\vec{p}),r}(G_n)} \leq c\Big(\sum_{j=0}^{\infty} 2^{sjr}\|u_j\|^r_{\mathrm{L}_{q,\vec{p}}(G_n)}\Big)^{\frac{1}{r}}.$$

With the help of Corollary 1.5.13 we can extend the previous lemma to the case of all integrability parameters being equal to infinity. Hence we get the following result.

Lemma 3.1.15. *Let $s \in \mathbb{R}$, $1 \leq r \leq \infty$. Then there exists a constant $c > 0$ such that any sequence $(u_j)_{j\in\mathbb{N}_0} \subset \mathscr{S}'(G_n)$ that fulfils the conditions (3.10) for some $A, B > 0$ and*

$$\Big(\sum_{j=0}^{\infty} 2^{sjr}\|u_j\|^r_{\mathrm{L}_{\infty}(G_n)}\Big)^{\frac{1}{r}} < \infty$$

the series $\sum_j u_j$ converges in $\mathscr{S}'(G_n)$ and in $\mathrm{B}^{s,(b,\vec{a})}_{\infty,r}(G_n)$, and it holds

$$\Big\|\sum_{j=0}^{\infty} u_j\Big\|_{\mathrm{B}^{s,(b,\vec{a})}_{\infty,r}(G_n)} \leq c\Big(\sum_{j=0}^{\infty} 2^{sjr}\|u_j\|^r_{\mathrm{L}_{\infty}(G_n)}\Big)^{\frac{1}{r}}.$$

Proof: We first show that the series converges in $\mathscr{S}'(G_n)$. The ideas are similar to the proof of Lemma 3.1.11. So we keep the proof short and mainly focus on the differences. We take $s_1 = s - \varepsilon$ for some $\varepsilon > 0$ and prove that the series is a Cauchy sequence in $\mathrm{B}_{\infty,\infty}^{s_1,(b,\vec{a})}(G_n)$. By the same arguments as in the proof of Lemma 3.1.11, see (3.13), we deduce

$$\Big\| \sum_{k=l}^{m} u_k \Big\|_{\mathrm{B}_{\infty,\infty}^{s_1,(b,\vec{a})}(G_n)} \lesssim \sup_{j \in \mathbb{N}_0} \sum_{k=l}^{m} 2^{s_1 k} \| \mathscr{F}^{-1} \varphi_j \mathscr{F} u_k \|_{\mathrm{L}_\infty(G_n)}$$

$$= \sup_{j \in \mathbb{N}_0} \sum_{k=l}^{m} 2^{s_1 k} \| \mathscr{F}_{\mathbb{R}^{n+1}}^{-1} \psi_j \big(\tfrac{\mathcal{T}}{2\pi} \cdot, \cdot \big) \mathscr{F}_{\mathbb{R}^{n+1}} u_k \circ \pi \|_{\mathrm{L}_\infty(\mathbb{R}^{n+1})},$$

where the last equality is due to Corollary 1.5.13 and φ_j is the restriction of a suitable cut-off function ψ_j by Definition 3.1.2. The functions $\psi_j(\tfrac{\mathcal{T}}{2\pi} \cdot, \cdot)$ still satisfy the requirements of Lemma 3.1.9 and hence inequality (3.8) is applicable. The continuity of the strong maximal operator in $\mathrm{L}_\infty(\mathbb{R}^{n+1})$ together with (3.8) yields the estimate

$$\Big\| \sum_{k=n}^{m} u_k \Big\|_{\mathrm{B}_{\infty,\infty}^{s_1,(b,\vec{a})}(G_n)} \lesssim \sum_{k=n}^{m} 2^{s_1 k} \| M_S[u_k \circ \pi] \|_{\mathrm{L}_\infty(\mathbb{R}^{n+1})} \lesssim \sum_{k=n}^{m} 2^{s_1 k} \| u_k \circ \pi \|_{\mathrm{L}_\infty(\mathbb{R}^{n+1})}$$

$$= \sum_{k=n}^{m} 2^{s_1 k} \| u_k \|_{\mathrm{L}_\infty(G_n)}.$$

Theorem 3.1.8 a) implies the embedding $\mathrm{B}_{\infty,r}^{s,(b,\vec{a})}(G_n) \hookrightarrow \mathrm{B}_{\infty,r}^{s_1,(b,\vec{a})}(G_n)$ since $s_1 < s$ and hence convergence of the last expression. Therefore, part d) of this theorem yields convergence in $\mathscr{S}'(G_n)$. Repeating the proof of Lemma 3.1.13 combined with Corollary 1.5.13 and using equation (3.8) instead of Lemma 3.1.9 concludes the result. $\qquad \square$

Remark 3.1.16. We started this section with the question of whether the defined spaces depend on the choice of functions φ_j or constants c, d. The previously shown result give an answer to this problem. If for any element $v \in \mathscr{S}'(G_n)$ the expression $F := \| \{ 2^{js} \mathscr{F}^{-1}[\lambda_j \mathscr{F} v] \}_{j \in \mathbb{N}_0} \|_{\mathrm{L}_{q,\vec{p}}(G_n, \ell_r)}$ is finite for any sequence $\{ \lambda_j \}_{j \in \mathbb{N}_0} \subset \mathscr{S}(\widehat{G}_n)$ that satisfies (3.10) and $\sum_{j=0}^{\infty} \lambda_j = 1$, and some parameters $1 < \vec{p}, q < \infty$, $s \in \mathbb{R}$ and $1 < r \le \infty$, then Lemma 3.1.13 implies $v \in \mathrm{F}_{(q,\vec{p}),r}^{s,(b,\vec{a})}(G_n)$ with a norm estimate $\| v \|_{\mathrm{F}_{(q,\vec{p}),r}^{s,(b,\vec{a})}(G_n)} \lesssim F$. Equivalence follows from the fact that if the spaces were defined by λ_j Lemma 3.1.13 can be proven with λ_j instead of φ_j and hence we obtain the opposite estimate and therefore equivalence of the definitions.

An important application of the independence lemmata is to show density of $\mathscr{S}(G_n)$ in Triebel-Lizorkin or Besov spaces. It is clear that we must exclude the cases of any index being infinity. Then we have the following result:

Lemma 3.1.17. *Let* $s \in \mathbb{R}$, $1 < \vec{p}, q < \infty$ *and* $1 < r < \infty$. *The space* $\mathscr{S}(G_n)$ *is dense in* $\mathrm{F}_{(q,\vec{p}),r}^{s,(b,\vec{a})}(G_n)$ *and in* $\mathrm{B}_{(q,\vec{p}),r}^{s,(b,\vec{a})}(G_n)$.

Proof: We start with $u \in \mathrm{F}_{(q,\vec{p}),r}^{s,(b,\vec{a})}(G_n)$. Since $u = \sum_{j=0}^{\infty} \mathscr{F}^{-1}[\varphi_j \mathscr{F} u]$ we get that the functions $u^N := \sum_{j=0}^{N} \mathscr{F}^{-1}[\varphi_j \mathscr{F} u]$ converge to u in $\mathrm{F}_{(q,\vec{p}),r}^{s,(b,\vec{a})}(G_n)$ by Lemma 3.1.13 for $N \to \infty$. By Theorem 1.5.11 we know that each distribution $\mathscr{F}^{-1}[\varphi_j \mathscr{F} u]$ is in $\mathrm{C}^\infty(G_n)$ with polynomial bound and therefore u^N as well. We define $\psi = c\varphi_0(0, \cdot) \in \mathrm{C}_0^\infty(\mathbb{R}^n)$, where $c > 0$ is chosen such that $[\mathscr{F}_{\mathbb{R}^n}^{-1} \psi](0) = 1$. This is possible as φ_0 is a positive function. So we get $u^N(\cdot)[\mathscr{F}_{\mathbb{R}^n}^{-1} \psi](\varepsilon \cdot) \in \mathscr{S}(G_n)$, the derivatives of u^N have a polynomial bound by Hörmander [46, Theorem 7.1.14] because derivatives directly apply to the exponential function. Furthermore it holds $u^N(\cdot)[\mathscr{F}_{\mathbb{R}^n}^{-1}\psi](\varepsilon\cdot) \to u^N(\cdot)$ in $L_{q,\vec{p}}(G_n)$ for $\varepsilon \to 0$. We have

$$\mathscr{F}[u^N(\cdot)[\mathscr{F}_{\mathbb{R}^n}^{-1}\psi](\varepsilon\cdot)] = \sum_{j=0}^{N} [c(\varepsilon)\varphi_j(\cdot)\mathscr{F}u(\cdot)] *_{\mathbb{R}^n} \psi\left(\frac{\cdot}{\varepsilon}\right),$$

and since $\psi\left(\frac{\cdot}{\varepsilon}\right)$ has support contained in $B_{\vec{a}}(0,2)$ uniformly for $\varepsilon < 1$, we conclude that

$$\operatorname{supp}\mathscr{F}[u^N(\cdot)\mathscr{F}^{-1}\psi(\varepsilon\cdot)] \subset B_{\vec{a}}(0,R)$$

for some $R > 0$ which depends only on N. With this we obtain

$$\|u^N - u^N \cdot [\mathscr{F}^{-1}\psi](\varepsilon\cdot)\|_{\mathrm{F}_{(q,\vec{p}),r}^{s,(b,\vec{a})}(G_n)} = \left\|\left(\sum_{k=0}^{\infty} 2^{ksr}|\mathscr{F}^{-1}[\varphi_k\mathscr{F}(u^N(\cdot) - u^N(\cdot)\mathscr{F}^{-1}\psi(\varepsilon\cdot))]|^r\right)^{\frac{1}{r}}\right\|_{L_{q,\vec{p}}}$$

$$\leq \left\|\sum_{k=0}^{M} 2^{ks}|\mathscr{F}^{-1}[\varphi_k\mathscr{F}(u^N(\cdot) - u^N(\cdot)\mathscr{F}^{-1}\psi(\varepsilon\cdot))]|\right\|_{L_{q,\vec{p}}}$$

for some constant $M = M(N)$ which is independent of ε. An application of Young's inequality yields

$$\|u^N(\cdot) - u^N(\cdot)\mathscr{F}^{-1}\psi(\varepsilon\cdot)\|_{\mathrm{F}_{(q,\vec{p}),r}^{s,(b,\vec{a})}(G_n)} \leq \sum_{k=0}^{M} 2^{ks}\|\mathscr{F}^{-1}\varphi_k\|_{L_1(G_n)}\|u^N(\cdot) - u^N(\cdot)\mathscr{F}^{-1}\psi(\varepsilon\cdot)\|_{L_{q,\vec{p}}(G_n)}$$

$$\lesssim \|u^N(\cdot) - u^N(\cdot)\mathscr{F}^{-1}\psi(\varepsilon\cdot)\|_{L_{q,\vec{p}}(G_n)} \to 0 \quad \text{for } \varepsilon \to 0.$$

As u^N converges to u in the norm of $\mathrm{F}_{(q,\vec{p}),r}^{s,(b,\vec{a})}(G_n)$ for $N \to \infty$, we obtain density of $\mathscr{S}(G_n)$ in $\mathrm{F}_{(q,\vec{p}),r}^{s,(b,\vec{a})}(G_n)$. The case of Besov spaces follows from the same arguments and the usage of Lemma 3.1.14 instead. $\qquad\square$

3.1.2 Embedding Results

In this section we investigate embedding results of Triebel-Lizorkin spaces and the question in what way the anisotropy influences regularity. We start with more consequences of the previous independence lemmata and show a result regarding differential operators, which will be of use in Section 3.4.

Lemma 3.1.18. *Let $s \in \mathbb{R}$, $1 < \vec{p}, q < \infty$ and $1 < r \leq \infty$. For every $\alpha \in \mathbb{N}_0^{n+1}$ such that $\alpha := (\alpha_0, \alpha_1, \ldots, \alpha_n)$ the differential operator $D^\alpha = D_t^{\alpha_0} D_{x_1}^{\alpha_1} \cdots D_{x_n}^{\alpha_n}$ is a continuous operator from $\mathrm{F}_{(q,\vec{p}),r}^{s,(b,\vec{a})}(G_n)$ to $\mathrm{F}_{(q,\vec{p}),r}^{s-\alpha\cdot(b,\vec{a}),(b,\vec{a})}(G_n)$.*

Proof: For any $u \in \mathscr{S}'(G_n)$ we have $u = \sum_{j=0}^{\infty} \mathscr{F}^{-1}[\varphi_j \mathscr{F}u]$. By Definition 3.1.2 we know that Ψ_j fulfils $\mathscr{F}^{-1}[\varphi_j\Psi_j\mathscr{F}u] = \mathscr{F}^{-1}[\varphi_j\mathscr{F}u]$ for every $j \in \mathbb{N}_0$. We split $\alpha = (\alpha_0, \alpha')$ with $\alpha_0 \in \mathbb{N}_0$ and $\alpha' \in \mathbb{N}_0^n$ to obtain

$$D^\alpha u = \sum_{j=0}^{\infty} \mathscr{F}^{-1}[c(\alpha)k^{\alpha_0}\xi^{\alpha'}\Psi_j\varphi_j\mathscr{F}u] = \sum_{j=0}^{\infty} \mathscr{F}^{-1}[c(\alpha)(2^{-jb}k)^{\alpha_0}(2^{-j\vec{a}}\xi)^{\alpha'}\Psi_j 2^{jb\alpha_0}2^{j\vec{a}\cdot\alpha'}\varphi_j\mathscr{F}u],$$

where the constant $c(\alpha)$ just contains i and $\frac{2\pi}{T}$ with the corresponding exponents. By applying Lemma 3.1.13, we derive

$$\|D^\alpha u\|_{F^{s-\alpha\cdot(b,\vec{a}),(b,\vec{a})}_{(q,\vec{p}),r}(G_n)} \lesssim \left\| \left(\sum_{j=0}^\infty 2^{jr(s-\alpha\cdot(b,\vec{a}))} |\mathscr{F}^{-1}[k^{\alpha_0}\xi^{\alpha'} \Psi_j \varphi_j \mathscr{F}u]|^r \right)^{\frac{1}{r}} \right\|_{L_{q,\vec{p}}(G_n)}$$

$$= \left\| \left(\sum_{j=0}^\infty 2^{jrs} |\mathscr{F}^{-1}[\tilde{\Psi}_j \varphi_j \mathscr{F}u]|^r \right)^{\frac{1}{r}} \right\|_{L_{q,\vec{p}}(G_n)},$$

with $\tilde{\Psi}_j(k,\xi) = (2^{-jb}k)^{\alpha_0}(2^{-j\vec{a}}\xi)^{\alpha'} \Psi_j(k,\xi)$. If we consider $\tilde{\Psi}_j$ as a function in $\mathbb{R}^{n+1}$, which is possible by removing the restriction of Ψ_j to $\hat{G}_n$ and replacing k with a continuous variable, it satisfies the requirements of Lemma 3.1.9. Hence applying Lemma 3.1.9 together with Theorem 2.2.5 yield

$$\|D^\alpha u\|_{F^{s-\alpha\cdot(b,\vec{a}),(b,\vec{a})}_{(q,\vec{p}),r}(G_n)} \lesssim \left\| \left(\sum_{j=0}^\infty 2^{jrs} |\mathscr{F}^{-1}[\varphi_j \mathscr{F}u]|^r \right)^{\frac{1}{r}} \right\|_{L_{q,\vec{p}}(G_n)} = \|u\|_{F^{s,(b,\vec{a})}_{(q,\vec{p}),r}(G_n)},$$

which completes the proof. $\qquad\qquad\qquad\qquad\qquad\qquad\qquad\qquad\qquad\qquad\qquad\qquad\qquad\square$

For the classical Triebel-Lizorkin spaces it is well known that the parameter s directly corresponds to differentiability of the function. In the anisotropic case the regularity is influenced by the parameter s and the anisotropy itself. Our goal is to show that the determining factor is the quotient of these two. For this reason we first extend Lemma 3.1.13 to allow for more options in the support of the functions u_j.

Lemma 3.1.19. *Let $s \in \mathbb{R}$, $\lambda > 0$, $1 < \vec{p}, q < \infty$, and $1 < r \le \infty$. Then there exists a constant $c > 0$ such that for every sequence $(u_j)_{j\in\mathbb{N}_0} \subset \mathscr{S}'(G_n)$ that fulfils $\operatorname{supp}\mathscr{F}u_0 \subset B_{(b,\vec{a})}(0,A)$,*

$$\operatorname{supp}\mathscr{F}u_j \subset \left\{ (k,\xi) \in \hat{G}_n \mid B2^{j\lambda} \le |(k,\xi)|_{b,\vec{a}} \le A2^{j\lambda} \right\}$$

for $j \ge 1$ and some constants $A, B \ge 0$, and

(3.16)
$$\left\| \left(\sum_{j=0}^\infty 2^{\lambda s jr} |u_j|^r \right)^{\frac{1}{r}} \right\|_{L_{q,\vec{p}}(G_n)} < \infty,$$

the series $\sum_{j=0}^\infty u_j$ converges in $\mathscr{S}'(G_n)$ and in $F^{s,(b,\vec{a})}_{(q,\vec{p}),r}(G_n)$ and it holds

$$\left\| \sum_{j=0}^\infty u_j \right\|_{F^{s,(b,\vec{a})}_{(q,\vec{p}),r}(G_n)} \le c \left\| \left(\sum_{j=0}^\infty 2^{\lambda s jr} |u_j|^r \right)^{\frac{1}{r}} \right\|_{L_{q,\vec{p}}(G_n)}.$$

Proof: Convergence in $\mathscr{S}'(G_n)$ follows from Lemma 3.1.11 by adopting the estimates from the beginning of the proof of Lemma 3.1.12.

We determine j, k such that $\varphi_k \mathscr{F}u_j = 0$ holds. This is the case if either $A2^{\lambda j} < 2^{k-1}$ or $3 \cdot 2^{k-1} < B2^{\lambda j}$. Hence there exists $\tilde{h} \in \lambda\mathbb{N}$ only dependent on A, B such that $\frac{k}{\lambda} - \frac{\tilde{h}}{\lambda} > j$ or $\frac{k}{\lambda} + \frac{\tilde{h}}{\lambda} < j$ implies $\varphi_k \mathscr{F}u_j = 0$. As we are interested in the corresponding integer values, we denote by $\lfloor x \rfloor$ the integer part of x and define $h = \frac{\tilde{h}}{\lambda} \in \mathbb{N}$ to derive $\lfloor \frac{k}{\lambda} \rfloor - h + 1 \ge j$ or $\lfloor \frac{k}{\lambda} \rfloor + h < j$ as necessary conditions

By construction we have that $\Psi_1(\xi) \neq 0$ only if $\frac{3}{4} \leq |\xi|_{(b,\vec{a})} \leq 4$. Since $|2^{(b,\vec{a})}x|_{(b,\vec{a})} = 2|x|_{(b,\vec{a})}$ the supremum above reduces to $x \in \mathbb{R}^{n+1}$ such that $\frac{1}{4} \leq \frac{3}{8} \leq |x|_{(b,\vec{a})} \leq 2 \leq 4$. All derivatives of Λ are bounded by (3.17) and together with $\Psi_1 \in \mathscr{S}(\mathbb{R}^{n+1})$ this yields the estimate

$$\sum_{|\alpha|,|\gamma|\leq N} 2^{-jl}\rho_{\alpha,0,\gamma}\left(\Psi_j(2^{j(b,\vec{a})}\cdot)\Lambda(2^{j(b,\vec{a})}\cdot)\right) \lesssim \sum_{|\alpha|\leq N} C_\alpha(\Lambda).$$

Applying the previous estimate together with Lemma 1.7.3 to the estimate from (3.18), we deduce

$$(3.19) \qquad \left\|\left\{2^{-jl}\mathscr{F}_{\mathbb{R}^{n+1}}^{-1}\Psi_j\Lambda\mathscr{F}_{\mathbb{R}^{n+1}}v_j\right\}_{j\in\mathbb{N}_0}\right\|_{L_{q,\vec{p}}(\mathbb{R}^{n+1},\ell_r)} \lesssim \sum_{|\alpha|\leq N} C_\alpha(\Lambda)\left\|\{v_j\}_{j\in\mathbb{N}_0}\right\|_{L_{q,\vec{p}}(\mathbb{R}^{n+1},\ell_r)}.$$

Since the operator $\Lambda(D)$ is well defined and $\varphi_j = \Phi_j\varphi_j$, we obtain for every $u \in \mathrm{F}^{s,(b,\vec{a})}_{(q,\vec{p}),r}(G_n)$

$$\|\Lambda(D)u\|_{\mathrm{F}^{s-l,(b,\vec{a})}_{(q,\vec{p}),r}(G_n)} = \left\|\left\{2^{j(s-l)}\mathscr{F}^{-1}[\Phi_j\varphi_j\Lambda|_{\widehat{G}_n}\mathscr{F}u]\right\}_{j\in\mathbb{N}_0}\right\|_{L_{q,\vec{p}}(G_n,\ell_r)}$$
$$\lesssim \left\|\left\{2^{js}\mathscr{F}^{-1}[\varphi_j\mathscr{F}u]\right\}_{j\in\mathbb{N}_0}\right\|_{L_{q,\vec{p}}(G_n,\ell_r)} = \|u\|_{\mathrm{F}^{s,(b,\vec{a})}_{(q,\vec{p}),r}(G_n)},$$

by estimate (3.19) together with Theorem 2.2.5 and $\mathcal{B}_1 = \mathcal{B}_2 = \ell_r$. $\qquad\square$

One simple application of the previous lemma is to show that the function $\langle\cdot\rangle^s_{b,\vec{a}}$, defined in Lemma 1.4.4, acts as a lift operator on the Triebel-Lizorkin scale.

Lemma 3.1.23. *The operator* $\Gamma_l : \mathscr{S}'(G_n) \to \mathscr{S}'(G_n)$ *given by* $\Gamma_l u := \mathscr{F}^{-1}[\langle(k,\xi)\rangle^l_{b,\vec{a}}\mathscr{F}u]$ *is a homeomorphism from* $\mathrm{F}^{s,(b,\vec{a})}_{(q,\vec{p}),r}(G_n) \to \mathrm{F}^{s-l,(b,\vec{a})}_{(q,\vec{p}),r}(G_n)$ *for every* $l \in \mathbb{R}$, $1 < q, \vec{p} < \infty$, *and* $1 < r \leq \infty$.

Proof: The idea is to apply Lemma 3.1.22 to $\Lambda(\xi) = \langle\xi\rangle^l_{b,\vec{a}}$ for $\xi \in \mathbb{R}^{n+1}$. Hence we need to estimate $C_\alpha(\Lambda)$ from equation (3.17) for all $\alpha \in \mathbb{N}_0^{n+1}$. We first consider the case of $l - (b,\vec{a}) \cdot \alpha \geq 0$. With the properties stated in Lemma 1.4.4 we derive

$$2^{-lj}\left|D^\alpha\left(\langle 2^{j(b,\vec{a})}\xi\rangle^l_{b,\vec{a}}\right)\right| \lesssim 2^{j(b,\vec{a})\cdot\alpha-lj}\langle 2^{j(b,\vec{a})}\xi\rangle^{l-(b,\vec{a})\cdot\alpha}_{b,\vec{a}} \leq \langle\xi\rangle^{l-(b,\vec{a})\cdot\alpha}_{b,\vec{a}}.$$

For $l - (b,\vec{a}) \cdot \alpha < 0$ we conclude the estimate

$$2^{-lj}\left|D^\alpha\left(\langle 2^{j(b,\vec{a})}\xi\rangle^l_{b,\vec{a}}\right)\right| \lesssim 2^{j(b,\vec{a})\cdot\alpha-lj}\langle 2^{j(b,\vec{a})}\xi\rangle^{l-(b,\vec{a})\cdot\alpha}_{b,\vec{a}} \leq |\xi|^{l-(b,\vec{a})\cdot\alpha}_{b,\vec{a}},$$

because $|\xi|_{(b,\vec{a})} \leq \langle\xi\rangle_{b,\vec{a}}$. Now boundedness of Γ_l follows from Lemma 3.1.22 since all terms on the right-hand side of the estimates are continuous on the compact set $\frac{1}{4} \leq |\xi|_{(b,\vec{a})} \leq 4$. As $\Gamma_l^{-1} = \Gamma_{-l}$, the statement follows. $\qquad\square$

Now we have all the tools needed to show the equivalence of spaces. Especially the previous lemma simplifies the proof of the first assertion to results from Section 2.3. These results are well known in isotropic case and can be found in Triebel [79, Section 2.3.1] for example.

Proposition 3.1.24. *For any* $1 < \vec{p}, q < \infty$ *and* $s \in \mathbb{R}$ *we have the following equalities of spaces with equivalent norms:*

a) $\mathrm{F}^{s,(b,\vec{a})}_{(q,\vec{p}),2}(G_n) = \mathrm{H}^{s,(b,\vec{a})}_{q,\vec{p}}(G_n)$.

b) *For* $l = \frac{s}{b} \in \mathbb{N}_0$ *and* $m_k = \frac{s}{a_k} \in \mathbb{N}_0$ *for every* $k \in \{1, 2, \ldots, n\}$ *we get* $\mathrm{F}^{s,(b,\vec{a})}_{(q,\vec{p}),2}(G_n) = \mathrm{W}^{l,\vec{m}}_{q,\vec{p}}(G_n)$.

Proof:

a) First we note that by Lemma 3.1.23 the $\mathrm{F}^{s,(b,\vec{a})}_{(q,\vec{p}),2}(G_n)$-norm of u is equivalent to the $\mathrm{F}^{0,(b,\vec{a})}_{(q,\vec{p}),2}(G_n)$-norm of $\mathscr{F}^{-1}[\langle(k,\xi)\rangle^s_{b,\vec{a}}\mathscr{F}u]$. Hence it suffices to consider $s = 0$, because Γ_l from this Lemma is obviously a lift for $\mathrm{H}^{s,(b,\vec{a})}_{q,\vec{p}}(G_n)$. Corollary 2.3.11 yields the equality of spaces and equivalence of norms in that case.

b) Let us first consider the functions

$$\Lambda_\alpha(\xi) := \frac{\xi^\alpha}{\langle\xi\rangle^s_{b,\vec{a}}}, \quad \xi \in \mathbb{R}^{n+1},$$

for $\alpha = 0$, $\alpha = le_1 =: m_0 e_1$ or $\alpha = m_k e_{k+1}$ for any $1 \le k \le n$. We show that these are $\mathrm{L}_{q,\vec{p}}(\mathbb{R}^{n+1})$-multipliers by Theorem 1.3.1. We start with $\alpha = le_1$ and consider $\gamma \in \mathbb{N}_0^{n+1}$ with $\gamma_j \le 1$ to derive

$$D^\gamma \Lambda_\alpha(\xi) = \delta_1(\gamma_1) l \xi^{\alpha-e_1} D^{\gamma-e_1} \langle\xi\rangle^{-s}_{b,\vec{a}} + \xi^\alpha D^\gamma \langle\xi\rangle^{-s}_{b,\vec{a}}.$$

Lemma 1.4.4 implies the estimate $|\xi^\gamma D^\gamma \langle\xi\rangle^s_{b,\vec{a}}| \lesssim \langle\xi\rangle^s_{b,\vec{a}}$ for any $s \in \mathbb{R}$. Hence we obtain

$$|\xi^\gamma D^\gamma \Lambda_\alpha(\xi)| \lesssim |\delta_1(\gamma_1)\xi^{\alpha+\gamma-e_1} D^{\gamma-e_1} \langle\xi\rangle^{-s}_{b,\vec{a}}| + |\xi^{\alpha+\gamma} D^\gamma \langle\xi\rangle^{-s}_{b,\vec{a}}|$$
$$\lesssim |\xi^\alpha \langle\xi\rangle^{-s}_{b,\vec{a}}| \lesssim 1.$$

The last estimate uses $|\xi^\alpha| \lesssim \langle\xi\rangle^{\alpha\cdot(b,\vec{a})}_{b,\vec{a}} = \langle\xi\rangle^s_{b,\vec{a}}$ which follows directly from Proposition 1.4.2 d). All other cases of α can be estimated in exactly the same way and therefore Theorem 1.3.1 together with Theorem 2.2.5 imply that $\Lambda_\alpha|_{\widehat{G_n}}$ are $\mathrm{L}_{q,\vec{p}}(G_n)$-multipliers. Hence for any $f \in \mathscr{S}(G_n)$ we have

$$\|f\|_{\mathrm{W}^{l,\vec{m}}_{q,\vec{p}}(G_n)} = \|f\|_{\mathrm{L}_{q,\vec{p}}(G_n)} + \|\partial^l_t f\|_{\mathrm{L}_{q,\vec{p}}(G_n)} + \sum_{j=1}^n \|\partial^{m_j}_j f\|_{\mathrm{L}_{q,\vec{p}}(G_n)}$$

$$= \|\mathscr{F}^{-1}\Lambda_0|_{\widehat{G_n}}\langle k,\xi\rangle^s_{b,\vec{a}}\mathscr{F}f\|_{\mathrm{L}_{q,\vec{p}}(G_n)} + \sum_{j=0}^n \|\mathscr{F}^{-1}\Lambda_{m_j e_{j+1}}|_{\widehat{G_n}}\langle k,\xi\rangle^s_{b,\vec{a}}\mathscr{F}f\|_{\mathrm{L}_{q,\vec{p}}(G_n)}$$

$$\lesssim \|\mathscr{F}^{-1}\langle k,\xi\rangle^s_{b,\vec{a}}\mathscr{F}f\|_{\mathrm{L}_{q,\vec{p}}(G_n)} + \sum_{j=0}^n \|\mathscr{F}^{-1}\langle k,\xi\rangle^s_{b,\vec{a}}\mathscr{F}f\|_{\mathrm{L}_{q,\vec{p}}(G_n)}$$

$$\lesssim \|f\|_{\mathrm{H}^{s,(b,\vec{a})}_{q,\vec{p}}(G_n)}.$$

By Lemma 3.1.17 and part a) the set $\mathscr{S}(G_n)$ is dense in $\mathrm{H}^{s,(b,\vec{a})}_{q,\vec{p}}(G_n)$. Therefore, we obtain $\|f\|_{\mathrm{W}^{l,\vec{m}}_{q,\vec{p}}(G_n)} \lesssim \|f\|_{\mathrm{H}^{s,(b,\vec{a})}_{q,\vec{p}}(G_n)}$ for any $f \in \mathrm{H}^{s,(b,\vec{a})}_{q,\vec{p}}(G_n) = \mathrm{F}^{s,(b,\vec{a})}_{(q,\vec{p}),2}(G_n)$.

For the opposite direction we take a $\lambda \in \mathrm{C}^\infty(\mathbb{R})$ such that $\lambda(-t) = -\lambda(t)$, and $\lambda = 1$ for $t \ge 1$. With this we define the function

$$\Theta(\xi) := \frac{\langle\xi\rangle^s_{b,\vec{a}}}{1 + \sum_{j=0}^n \lambda(\xi_{j+1})^{m_j}\xi^{m_j}_{j+1}}.$$

For the moment assume that Θ is an $L_{q,\vec{p}}(\mathbb{R}^{n+1})$-multiplier so that $\Theta|_{\widehat{G}_n}$ is an $L_{q,\vec{p}}(G_n)$-multiplier by Theorem 2.2.5. Hence for any $f \in \mathscr{S}(G_n)$ we have

$$\|f\|_{H^{s,(b,\vec{a})}_{(q,\vec{p})}(G_n)} \leq \|\mathscr{F}^{-1}\Theta|_{\widehat{G}_n}\mathscr{F}f\|_{L_{q,\vec{p}}(G_n)} + \|\mathscr{F}^{-1}\Theta|_{\widehat{G}_n}\lambda(k)^l k^l \mathscr{F}f\|_{L_{q,\vec{p}}(G_n)}$$

$$+ \sum_{j=1}^{n}\|\mathscr{F}^{-1}\Theta|_{\widehat{G}_n}\lambda(\xi_j)^{m_j}\xi_j^{m_j}\mathscr{F}f\|_{L_{q,\vec{p}}(G_n)}$$

$$\lesssim \|f\|_{L_{q,\vec{p}}(G_n)} + \|\mathscr{F}^{-1}\lambda(k)^l \mathscr{F}\partial_t^l f\|_{L_{q,\vec{p}}(G_n)} + \sum_{j=1}^{n}\|\mathscr{F}^{-1}\lambda(\xi_j)^{m_j}\mathscr{F}\partial_j^{m_j}f\|_{L_{q,\vec{p}}(G_n)}$$

$$\lesssim \|f\|_{W^{l,\vec{m}}_{q,\vec{p}}(G_n)},$$

as λ is an $L_{q,\vec{p}}(\mathbb{R}^n)$-multiplier since it is a smooth bounded function with every derivative vanishing outside a compact set. From Lemma 1.3.6 we know that $\mathscr{S}(G_n)$ is dense in $W^{l,\vec{m}}_{q,\vec{p}}(G_n)$ and therefore we obtain $\|f\|_{H^{s,(b,\vec{a})}_{(q,\vec{p})}(G_n)} \lesssim \|f\|_{W^{l,\vec{m}}_{q,\vec{p}}(G_n)}$ for any $f \in W^{l,\vec{m}}_{q,\vec{p}}(G_n)$.

So it remains to show that Θ is indeed a multiplier. To this aim we apply Theorem 1.3.1 and consider $\alpha \in \{0,1\}^{n+1}$. By Lemma 5.1.1 we derive

$$|\xi^\alpha D^\alpha \Theta(\xi)| \leq \sum_{k=0}^{|\alpha|}\sum_{\substack{\gamma \in \mathbb{N}_0^{n+1}, \\ \gamma \leq \alpha, \\ |\gamma|=k}} k! \frac{|\xi^{\alpha-\gamma}D^{\alpha-\gamma}\langle\xi\rangle^s_{b,\vec{a}}|\prod_{j=0}^{n}\left|\xi_{j+1}^{\gamma_{j+1}}\partial_{j+1}[\lambda(\xi_{j+1})^{m_j}\xi_{j+1}^{m_j}]\right|^{\gamma_{j+1}}}{(1+\sum_{j=0}^{n}\lambda(\xi_{j+1})^{m_j}\xi_{j+1}^{m_j})^{1+k}}$$

$$\lesssim \sum_{k=0}^{|\alpha|}\sum_{\substack{\gamma \in \mathbb{N}_0^{n+1}, \\ \gamma \leq \alpha, \\ |\gamma|=k}} \frac{\langle\xi\rangle^s_{b,\vec{a}}\prod_{j=0}^{n}\left(|\xi_{j+1}^{m_j}| + |\xi_{j+1}^{m_j+1}\lambda'(\xi_{j+1})|\right)^{\gamma_{j+1}}}{(1+\sum_{j=0}^{n}\lambda(\xi_{j+1})^{m_j}\xi_{j+1}^{m_j})^{1+k}}.$$

Here we used Lemma 1.4.4 and the boundedness of λ. Proposition 1.4.2 yields $\langle\xi\rangle_{b,\vec{a}} = |(1,b,\vec{a})|^s_{1,b,\vec{a}} \lesssim 1 + \sum_{j=0}^{n}|\xi_{j+1}|^{m_j}$. By the properties of λ we conclude

$$(3.20) \qquad\qquad (\lambda(\xi_{j+1})\xi_{j+1})^{m_j} = |\xi_{j+1}|^{m_j}$$

for $|\xi_{j+1}| \geq 1$ and all $j = 0,1,2,\dots,n$. We only need to consider the terms without λ' since the derivative vanishes outside a compact set. For fixed k the product $\prod_{j=1}^{n}$ has exactly k non-constant factors as only k of the γ_j are not zero. The quotient of each of these factors together with one of the factors of the denominator is bounded by (3.20). By Theorem 1.3.1 and Theorem 2.2.5 we conclude that Θ is an $L_{q,\vec{p}}(G_n)$-multiplier and hence the proof is finished. $\qquad\square$

For estimating non-linear terms at the end of Chapter 4 we need embedding results which we provide here. Note that with our previous work their proofs are rather easy and quite straightforward.

Lemma 3.1.25. *Let $s,l \in \mathbb{R}$, $1 \leq \vec{p} \leq \vec{r} < \infty$ and $1 \leq q_1 \leq q_2 < \infty$ such that*

$$(3.21) \qquad\qquad s - \frac{b}{q_1} - \frac{\vec{a}}{\vec{p}} = l - \frac{b}{q_2} - \frac{\vec{a}}{\vec{r}} = l - \frac{b}{q_2} + \sum_{j=1}^{n}\frac{a_j}{r_j}$$

and $s > l$. Then the continuous embedding $F^{s,(b,\vec{a})}_{(q_1,\vec{p}),\infty}(G_n) \hookrightarrow F^{l,(b,\vec{a})}_{(q_2,\vec{r}),m}(G_n)$ holds for all m with $1 \leq m \leq \infty$.

Proof: As the support of φ_j is contained in $|(k,\xi)|_{b,\vec{a}} \le 2^{j+1}$, we obtain $|k| \le 2^{jb}2^b$ and $|\xi_i| \le 2^{ja_i}2^{a_i}$ for all $(k,\xi) \in \operatorname{supp}\varphi_j$. We choose $A = 2^{\max\{a_1,a_2,\ldots,a_n\}}$, $B = 2^b$, $K = 2^b$, $R_j = 2^{a_j}$ and apply Theorem 2.1.2 to $2^{jl}\mathscr{F}_{G_n}^{-1}[\varphi_j\mathscr{F}_{G_n}f]$. This yields

$$\|u\|_{\mathrm{F}^{l,(b,\vec{a})}_{(q_2,\vec{r}),m}(G_n)} = \left\|\left(\sum_{j=0}^{\infty} 2^{jlm}\left|\mathscr{F}^{-1}[\varphi_j\mathscr{F}u]\right|^m\right)^{\frac{1}{m}}\right\|_{\mathrm{L}_{q_2,\vec{r}}(G_n)}$$

$$\lesssim \left\|\sup_{j\in\mathbb{N}_0} 2^{jl}2^{\frac{jb}{q_1}-\frac{jb}{q_2}}\prod_{k=1}^{n} 2^{\frac{ja_k}{p_k}-\frac{ja_k}{r_k}}\left|\mathscr{F}^{-1}[\varphi_j\mathscr{F}u]\right|\right\|_{\mathrm{L}_{q_1,\vec{p}}(G_n)}$$

$$= \left\|\sup_{j\in\mathbb{N}_0} 2^{js}\left|\mathscr{F}^{-1}[\varphi_j\mathscr{F}u]\right|\right\|_{\mathrm{L}_{q_1,\vec{p}}(G_n)} = \|u\|_{\mathrm{F}^{s,(b,\vec{a})}_{(q_1,\vec{p}),\infty}(G_n)},$$

which proves the assertion. $\qquad\square$

Lemma 3.1.26. *Let $s,l \in \mathbb{R}$, $1 \le \vec{p} \le \vec{r} \le \infty$ and $1 \le q_1 \le q_2 \le \infty$ such that*

$$(3.22) \qquad s - \frac{b}{q_1} - \frac{\vec{a}}{\vec{p}} = l - \frac{b}{q_2} - \frac{\vec{a}}{\vec{r}}.$$

Then the continuous embedding $\mathrm{B}^{s,(b,\vec{a})}_{(q_1,\vec{p}),m}(G_n) \hookrightarrow \mathrm{B}^{l,(b,\vec{a})}_{(q_2,\vec{r}),m}(G_n)$ holds for all $m \in [1,\infty]$.

Proof: The approach is the same as in the proof of Lemma 3.1.25, but instead we apply Lemma 2.1.1 and obtain

$$\|u\|_{\mathrm{B}^{l,(b,\vec{a})}_{(q_2,\vec{r}),m}(G_n)} = \left(\sum_{j=0}^{\infty} 2^{jlm}\left\|\mathscr{F}^{-1}[\varphi_j\mathscr{F}u]\right\|^m_{\mathrm{L}_{q_2,\vec{r}}(G_n)}\right)^{\frac{1}{m}}$$

$$\lesssim \left(\sum_{j=0}^{\infty} 2^{jlm}2^{\frac{jmb}{q_1}-\frac{jmb}{q_2}}\prod_{k=1}^{n} 2^{\frac{jma_k}{p_k}-\frac{jma_k}{r_k}}\left\|\mathscr{F}^{-1}[\varphi_j\mathscr{F}u]\right\|^m_{\mathrm{L}_{q_1,\vec{p}}(G_n)}\right)^{\frac{1}{m}}$$

$$= \left(\sum_{j=0}^{\infty} 2^{jsm}\left\|\mathscr{F}^{-1}[\varphi_j\mathscr{F}u]\right\|^m_{\mathrm{L}_{q_1,\vec{p}}(G_n)}\right)^{\frac{1}{m}} = \|u\|_{\mathrm{B}^{s,(b,\vec{a})}_{(q_1,\vec{p}),m}(G_n)}.$$

For $m = \infty$ we exchange the sum for the supremum and repeat the steps to obtain the result. $\quad\square$

With the previous two lemmata we can further show embeddings into the space of continuous functions.

Corollary 3.1.27. *Let $1 \le \vec{p}, q < \infty$ and $1 \le r \le \infty$. For $s > \frac{b}{q} + \frac{\vec{a}}{\vec{p}}$ the continuous embedding $\mathrm{F}^{s,(b,\vec{a})}_{(q,\vec{p}),r}(G_n) \hookrightarrow \mathrm{C}(G_n)$ is valid.*

Proof: By assumption we find $\varepsilon > 0$ such that

$$\mathrm{F}^{s,(b,\vec{a})}_{(q,\vec{p}),r}(G_n) \hookrightarrow \mathrm{F}^{s,(b,\vec{a})}_{(q,\vec{p}),\infty}(G_n) \hookrightarrow \mathrm{B}^{s,(b,\vec{a})}_{(q,\vec{p}),\infty}(G_n) \hookrightarrow \mathrm{B}^{\varepsilon,(b,\vec{a})}_{\infty,\infty}(G_n) \hookrightarrow \mathrm{B}^{0,(b,\vec{a})}_{\infty,1}(G_n),$$

where we used Theorem 3.1.8 and Lemma 3.1.26. Because $u_j = \mathscr{F}^{-1}\varphi_j\mathscr{F}u$ is a continuous function by Theorem 1.5.11 for every $j \in \mathbb{N}_0$, we derive the continuous embedding from the fact that the series $\sum u_j$ converges in L_∞ by converging in $\mathrm{B}^{0,(b,\vec{a})}_{\infty,1}(G_n)$. $\qquad\square$

Proposition 3.1.32. *Let $1 < p, q < \infty$ and $s \in [0, \infty)$ such that $l = \frac{s}{b} \in \mathbb{N}_0$ and $m_k = \frac{s}{a_k} \in \mathbb{N}_0$ for every $k \in \{1, 2, \ldots, n\}$. The identity $\mathrm{F}^{s,(b,\vec{a})}_{(q,p),2}(\Omega_\mathbb{T}) = \mathrm{W}^{l,\vec{m}}_{q,p}(\Omega_\mathbb{T})$ holds with equivalent norms.*

Proof: First we consider $u \in \mathrm{F}^{s,(b,\vec{a})}_{(q,p),r}(\Omega_\mathbb{T})$. By definition we find $v \in \mathrm{F}^{s,(b,\vec{a})}_{(q,p),r}(G_n)$ such that $v = u$ on $\Omega_\mathbb{T}$ and

$$\|v\|_{\mathrm{F}^{s,(b,\vec{a})}_{(q,p),r}(G_n)} \leq 2\|u\|_{\mathrm{F}^{s,(b,\vec{a})}_{(q,p),r}(\Omega_\mathbb{T})}.$$

By Proposition 3.1.24 we have $v \in \mathrm{W}^{l,\vec{m}}_{q,p}(G_n)$. This implies $u \in \mathrm{W}^{l,\vec{m}}_{q,p}(\Omega_\mathbb{T})$ because $v = u$ on $\Omega_\mathbb{T}$ and it holds

$$\|u\|_{\mathrm{W}^{l,\vec{m}}_{q,p}(\Omega_\mathbb{T})} = \|v\|_{\mathrm{W}^{l,\vec{m}}_{q,p}(\Omega_\mathbb{T})} \leq \|v\|_{\mathrm{W}^{l,\vec{m}}_{q,p}(G_n)} \lesssim \|v\|_{\mathrm{F}^{s,(b,\vec{a})}_{(q,p),r}(\Omega_\mathbb{T})} \leq 2\|u\|_{\mathrm{F}^{s,(b,\vec{a})}_{(q,p),r}(\Omega_\mathbb{T})}.$$

The estimate of the Sobolev norm by the Triebel-Lizorkin norm follows from Proposition 3.1.24 with a constant independent of u and v.

Next let $u \in \mathrm{W}^{l,\vec{m}}_{q,p}(\Omega_\mathbb{T})$. By the Extension Theorem of Stein, see Stein [77, Chapter 6], or Adams and Fournier [1, Chapter 5] for a simpler approach, we find an extension Eu to u such that $Eu \in \mathrm{W}^{l,\vec{m}}_{q,p}(G_n)$ with $\|Eu\|_{\mathrm{W}^{l,\vec{m}}_{q,p}(G_n)} \lesssim \|u\|_{\mathrm{W}^{l,\vec{m}}_{q,p}(\mathbb{T}\times\Omega)}$; this part requires the regularity of Ω. By Proposition 3.1.24 we conclude $Eu \in \mathrm{F}^{s,(b,\vec{a})}_{(q,p),r}(G_n)$. Since Eu is an extension to u in $\mathrm{F}^{s,(b,\vec{a})}_{(q,p),r}(G_n)$, we obtain $u \in \mathrm{F}^{s,(b,\vec{a})}_{(q,p),r}(\Omega_\mathbb{T})$ and the norm estimate follows trivially. $\qquad\square$

Lemma 3.1.33. *Let $s \in \mathbb{R}$, $1 < q, p < \infty$, and $1 < r \leq \infty$. For every $\alpha \in \mathbb{N}_0^{n+1}$ such that $\alpha = (\alpha_0, \alpha_1, \ldots, \alpha_n)$ the differential operator $D^\alpha = D_t^{\alpha_0} D_{x_1}^{\alpha_1} \cdots D_{x_n}^{\alpha_n}$ is a continuous operator from $\mathrm{F}^{s,(b,\vec{a})}_{(q,p),r}(\Omega_\mathbb{T})$ to $\mathrm{F}^{s-\alpha\cdot(b,\vec{a}),(b,\vec{a})}_{(q,p),r}(\Omega_\mathbb{T})$.*

Proof: Lemma 3.1.18 shows that $D^\alpha v \in \mathrm{F}^{s-\alpha\cdot(b,\vec{a}),(b,\vec{a})}_{(q,p),r}(G_n)$ for any $v \in \mathrm{F}^{s,(b,\vec{a})}_{(q,p),r}(G_n)$. Hence it holds for $u \in \mathrm{F}^{s,(b,\vec{a})}_{(q,p),r}(\Omega_\mathbb{T})$

$$\begin{aligned}
\|D^\alpha u\|_{\mathrm{F}^{s-\alpha\cdot(b,\vec{a}),(b,\vec{a})}_{(q,p),r}(\Omega_\mathbb{T})} &= \inf\left\{\|v\|_{\mathrm{F}^{s-\alpha\cdot(b,\vec{a}),(b,\vec{a})}_{(q,p),r}(G_n)} \mid v \in \mathrm{F}^{s-\alpha\cdot(b,\vec{a}),(b,\vec{a})}_{(q,p),r}(G_n) \text{ with } D^\alpha u = v \text{ on } \Omega_\mathbb{T}\right\} \\
&\lesssim \inf\left\{\|w\|_{\mathrm{F}^{s,(b,\vec{a})}_{(q,p),r}(G_n)} \mid w \in \mathrm{F}^{s,(b,\vec{a})}_{(q,p),r}(G_n) \text{ such that } u = w \text{ on } \Omega_\mathbb{T}\right\} \\
&= \|u\|_{\mathrm{F}^{s,(b,\vec{a})}_{(q,p),r}(\Omega_\mathbb{T})},
\end{aligned}$$

which proves the assertion. $\qquad\square$

The next two results can be proven in a more general framework, but they are only formulated in the setting of Sobolev spaces to ease the applicability in Chapter 4.

Lemma 3.1.34. *Let $1 < p, q < \infty$ and $s \geq 0$ such that $l = \frac{s}{b} \in \mathbb{N}_0$ and $m_k = \frac{s}{a_k} \in \mathbb{N}_0$ for every $k \in \{1, 2, \ldots, n\}$. Then for $\kappa = (1-\theta)s$ with $\theta \in (0,1)$ we have*

$$\|u\|_{\mathrm{F}^{\kappa,(b,\vec{a})}_{(q,p),2}(\Omega_\mathbb{T})} \lesssim \|u\|_{\mathrm{L}_{q,p}(\Omega_\mathbb{T})}^\theta \|u\|_{\mathrm{W}^{l,\vec{m}}_{q,p}(\Omega_\mathbb{T})}^{1-\theta}$$

for all $u \in \mathrm{W}^{l,\vec{m}}_{q,p}(\Omega_\mathbb{T})$.

Proof: Let Eu be the same extension of u as in the proof of Proposition 3.1.32. Lemma 3.1.29 yields

$$\|u\|_{\mathrm{F}^{\kappa,(b,\bar{a})}_{(q,p),2}(\Omega_{\mathrm{T}})} \leq \|Eu\|_{\mathrm{F}^{\kappa,(b,\bar{a})}_{(q,p),2}(G_n)} \lesssim \|Eu\|^{\theta}_{\mathrm{F}^{0,(b,\bar{a})}_{(q,p),\infty}(G_n)} \|Eu\|^{1-\theta}_{\mathrm{F}^{s,(b,\bar{a})}_{(q,p),\infty}(G_n)}$$

$$\lesssim \|Eu\|^{\theta}_{\mathrm{L}_{q,p}(G_n)} \|Eu\|^{1-\theta}_{\mathrm{W}^{l,\bar{m}}_{q,p}(G_n)} \lesssim \|u\|^{\theta}_{\mathrm{L}_{q,p}(\Omega_{\mathrm{T}})} \|u\|^{1-\theta}_{\mathrm{W}^{l,\bar{m}}_{q,p}(\Omega_{\mathrm{T}})},$$

where we used the embeddings of Theorem 3.1.8 a) together with Proposition 3.1.24 a). $\qquad\square$

The last corollary of this section states further results in the setting of Sobolev spaces on domains, which are used for the estimates in Section 4.4. For $p = q$ the result can be found in Galdi and Kyed [38, Chapter 4].

Corollary 3.1.35. *Let $1 < q, p < \infty$. Then the estimate*

$$\|u\|_{\mathrm{L}_{q_0,p_0}(\Omega_{\mathrm{T}})} + \|\nabla u\|_{\mathrm{L}_{q_1,p_1}(\Omega_{\mathrm{T}})} \lesssim \|u\|_{\mathrm{W}^{1,2}_{q,p}(\Omega_{\mathrm{T}})}$$

holds for all $u \in \mathrm{W}^{1,2}_{q,p}(\Omega_{\mathrm{T}})$ and all $q_0, q_1 \in [q, \infty]$ and $p_0, p_1 \in [p, \infty]$ satisfying

$$\begin{cases} q_0 \leq \frac{2q}{2-\alpha q} & \text{if } \alpha q < 2, \\ q_0 < \infty & \text{if } \alpha q = 2, \\ q_0 \leq \infty & \text{if } \alpha q > 2, \end{cases} \qquad \begin{cases} p_0 \leq \frac{np}{n-(2-\alpha)p} & \text{if } (2-\alpha)p < n, \\ p_0 < \infty & \text{if } (2-\alpha)p = n, \\ p_0 \leq \infty & \text{if } (2-\alpha)p > n, \end{cases}$$

with $\alpha \in [0, 2]$ and

$$\begin{cases} q_1 \leq \frac{2q}{2-\beta q} & \text{if } \beta q < 2, \\ q_1 < \infty & \text{if } \beta q = 2, \\ q_1 \leq \infty & \text{if } \beta q > 2, \end{cases} \qquad \begin{cases} p_1 \leq \frac{np}{n-(1-\beta)p} & \text{if } (1-\beta)p < n, \\ p_1 < \infty & \text{if } (1-\beta)p = n, \\ p_1 \leq \infty & \text{if } (1-\beta)p > n, \end{cases}$$

with $\beta \in [0, 1]$.

Proof: By Proposition 3.1.32 we know that there is an extension $Eu \in \mathrm{F}^{2,(2,\vec{1})}_{q,p}(G_n)$ of u. Hence Lemma 3.1.28 yields the result with $\gamma^0_0 = \alpha$ and $\gamma^0_j = \frac{2-\alpha}{n}$ as well as $\gamma^{e_k}_0 = \beta$ and $\gamma^{e_k}_j = \frac{1-\beta}{n}$ for $k = 2, \ldots, n, n+1$. $\qquad\square$

3.2 Pointwise Multiplication

The main problem of pointwise multiplication of elements in Besov or Triebel-Lizorkin spaces is that in general the product $u \cdot v$ for $u, v \in \mathscr{S}'(G_n)$ is not well-defined. So the idea is to regularize both elements simultaneously and show convergence in the appropriate space. We know from Lemma 1.6.2 that for $u \in \mathscr{S}'(G_n)$ the sequence $\{u_m\}_{m \in \mathbb{N}}$ defined by $u_m := \sum_{j=0}^{m} \mathscr{F}^{-1} \varphi_j \mathscr{F} u$ converges to u in $\mathscr{S}'(G_n)$. By Theorem 1.5.11 every u_m is a smooth function and hence the product $u_m \cdot v_m$ is a well-defined function and element of $\mathscr{S}'(G_n)$. We follow the ideas of Johnsen [51] in extending the existing theory to the time-periodic framework. However, we make an adjustment to the functions ψ that are admissible in the definition to guarantee the properties of Lemma 3.2.5 below. Let $\psi \in \mathrm{C}^{\infty}_0(\mathbb{R}^{n+1})$ and $0 < R_2 < R_1$ such that $2R_2 > R_1$ and ψ satisfies

(3.23) $$\psi(x) = 1 \quad \text{for all } x \in B_{b,\bar{a}}(0, R_2), \quad \mathrm{supp}\, \psi \subset B_{b,\bar{a}}(0, R_1).$$

Definition 3.2.1. Let $u, v \in \mathscr{S}'(G_n)$ and let $\psi \in C_0^\infty(\mathbb{R}^{n+1})$ satisfy (3.23). Define $\mu_\psi(u, v)$ by

$$(3.24) \qquad \mu_\psi(u, v) := \lim_{m \to \infty} \mathscr{F}^{-1}[\psi(2^{-m(b,\vec{a})} \cdot)\big|_{\widehat{G}_n} \mathscr{F} u] \cdot \mathscr{F}^{-1}[\psi(2^{-m(b,\vec{a})} \cdot)\big|_{\widehat{G}_n} \mathscr{F} v]$$

whenever the limit exists in $\mathscr{S}'(G_n)$. The operator $\mu(u, v)$ is defined as $\mu(u, v) := \mu_\psi(u, v)$ if μ_ψ exists for all ψ satisfying (3.23) and consequently is independent of ψ. Furthermore, to simplify the notation we denote

$$\psi^m := \psi(2^{-m(b,\vec{a})} \cdot)\big|_{\widehat{G}_n}, \quad \varphi^m := \mathscr{F}^{-1}[\psi^m \mathscr{F} \varphi]$$

for $\varphi \in \mathscr{S}(G_n)$. For $u \in \mathscr{S}'(G_n)$ we define u^m in the same way as φ^m.

The restriction of ψ to $\widehat{G}_n$ is necessary in this case as it is not possible to simply scale a function in $C_0^\infty(\widehat{G}_n)$. In the following Lemma we show that u^m approximates u in the norm of Triebel-Lizorkin and Besov spaces.

Lemma 3.2.2. *Let* $1 < \vec{p}, q < \infty$, $1 < r < \infty$, $s \in \mathbb{R}$, $u \in F_{(q,\vec{p}),r}^{s,(b,\vec{a})}(G_n)$ *and* $v \in B_{(q,\vec{p}),r}^{s,(b,\vec{a})}(G_n)$. *Then* u^m *converges to* u *in* $F_{(q,\vec{p}),r}^{s,(b,\vec{a})}(G_n)$ *and* v^m *converges to* v *in* $B_{(q,\vec{p}),r}^{s,(b,\vec{a})}(G_n)$ *as* $m \to \infty$.

Proof: We first show the result in the case of $F_{(q,\vec{p}),r}^{s,(b,\vec{a})}(G_n)$. We note that there exists an $h \in \mathbb{N}$ such that $(1 - \psi^m)\varphi_j \neq 0$ only if $j \geq m - h$ and $\psi^m \varphi_j \neq 0$ only if $j \leq m + h$. With this we obtain

$$\|u - u^m\|_{F_{(q,\vec{p}),r}^{s,(b,\vec{a})}(G_n)} = \left\| \left(\sum_{j=0}^{\infty} 2^{jsr} |\mathscr{F}^{-1}(1 - \psi^m)\varphi_j \mathscr{F} u|^r \right)^{\frac{1}{r}} \right\|_{L_{q,\vec{p}}(G_n)}$$

$$\leq \left\| \left(\sum_{j=m-h}^{\infty} 2^{jsr} |\mathscr{F}^{-1}\varphi_j \mathscr{F} u|^r \right)^{\frac{1}{r}} \right\|_{L_{q,\vec{p}}(G_n)} + \left\| \left(\sum_{j=m-h}^{m+h} 2^{jsr} |\mathscr{F}^{-1}\psi^m \varphi_j \mathscr{F} u|^r \right)^{\frac{1}{r}} \right\|_{L_{q,\vec{p}}(G_n)}.$$

The first term converges to 0 for $m \to \infty$, hence we only consider the second term. The monotonicity of ℓ_r spaces and Lemma 3.1.9 together with Theorem 2.2.5 applied to ψ^m instead of ψ_j yield

$$\left\| \left(\sum_{j=m-h}^{m+h} 2^{jsr} |\mathscr{F}^{-1}\psi^m \varphi_j \mathscr{F} u|^r \right)^{\frac{1}{r}} \right\|_{L_{q,\vec{p}}(G_n)} \leq \sum_{j=m-h}^{m+h} \left\| 2^{js} \mathscr{F}^{-1}\psi^m \varphi_j \mathscr{F} u \right\|_{L_{q,\vec{p}}(G_n)}$$

$$\lesssim \sum_{j=m-h}^{m+h} \left\| 2^{js} \mathscr{F}^{-1}\varphi_j \mathscr{F} u \right\|_{L_{q,\vec{p}}(G_n)}$$

$$\lesssim (2h + 1) \left\| \left(\sum_{j=m-h}^{\infty} 2^{jsr} |\mathscr{F}^{-1}\varphi_j \mathscr{F} u|^r \right)^{\frac{1}{r}} \right\|_{L_{q,\vec{p}}(G_n)}.$$

The last term converges to 0 by the dominated convergence theorem and hence we derive convergence of u^m to u in $F_{(q,\vec{p}),r}^{s,(b,\vec{a})}(G_n)$. The convergence in Besov spaces follows by the same arguments. $\qquad \square$

In the same way that we extended the independence result for Besov spaces to the integrability being infinity we are going to extend Lemma 3.2.2. At the end of this section we will see the necessity of this.

Lemma 3.2.3. *Let $s \in \mathbb{R}$, $1 \le r < \infty$, and $v \in \mathrm{B}^{s,(b,\bar{a})}_{\infty,r}(G_n)$. Then v^m converges to v in $\mathrm{B}^{s,(b,\bar{a})}_{\infty,r}(G_n)$ as $m \to \infty$.*

Proof: We take h with the same properties as in Lemma 3.2.2. To make use of Corollary 1.5.13 we define $\Theta^m := \psi(2^{-m(b,\bar{a})}\cdot)$ and recall that ψ_j is the unrestricted version of φ_j according to Definition 3.1.2. It holds

$$\|v - v^m\|^r_{\mathrm{B}^{s,(b,\bar{a})}_{\infty,r}(G_n)} = \sum_{j=m-h}^{\infty} 2^{jsr}\|\mathscr{F}^{-1}(1-\psi^m)\varphi_j\mathscr{F}v\|^r_{\mathrm{L}_\infty(G_n)}$$

$$\le \sum_{j=m-h}^{\infty} 2^{jsr}\|\mathscr{F}^{-1}\varphi_j\mathscr{F}v\|^r_{\mathrm{L}_\infty(G_n)} + \sum_{j=m-h}^{\infty} 2^{jsr}\|\mathscr{F}^{-1}\psi^m\varphi_j\mathscr{F}v\|^r_{\mathrm{L}_\infty(G_n)}$$

Since the first term converges to 0 we examine the second. Let $\omega = \frac{\mathcal{I}}{2\pi}$ and by applying (3.9) with the absolute value instead of an ℓ_q-norm, Θ^m for ψ_j and together with the L_∞-estimate of the maximal operator we obtain

$$\sum_{j=m-h}^{\infty} 2^{jsr}\|\mathscr{F}^{-1}\psi^m\varphi_j\mathscr{F}v\|^r_{\mathrm{L}_\infty(G_n)} = \sum_{j=m-h}^{\infty} 2^{jsr}\|\mathscr{F}^{-1}_{\mathbb{R}^{n+1}}\big[(\Theta^m\psi_j)(\omega\cdot,\cdot)\mathscr{F}_{\mathbb{R}^{n+1}}v \circ \pi\big]\|^r_{\mathrm{L}_\infty(\mathbb{R}^{n+1})}$$

$$\lesssim \sum_{j=m-h}^{\infty} 2^{jsr}\|\mathscr{F}^{-1}_{\mathbb{R}^{n+1}}\big[\psi_j(\omega\cdot,\cdot)\mathscr{F}_{\mathbb{R}^{n+1}}v \circ \pi\big]\|^r_{\mathrm{L}_\infty(\mathbb{R}^{n+1})}$$

$$\lesssim \sum_{j=m-h}^{\infty} 2^{jsr}\|\mathscr{F}^{-1}\varphi_j\mathscr{F}v\|^r_{\mathrm{L}_\infty(G_n)}.$$

The last term converges to 0 as $m \to \infty$ and we therefore get convergence. $\qquad\square$

As a next step we introduce operators called paraproducts or paramultiplication operators. In the framework of $\mathbb{R}^n$, these operators have been investigated for example in Runst and Sickel [72], Johnsen [51] and Yamazaki [85,86]. The last two authors considered them in an anisotropic setting, and all of them for integrability given by $p \in (1, \infty)$.

Definition 3.2.4. Let ψ be as in Definition 3.2.1 and $\psi_j = \psi^j = 0$ for $j < 0$. Set $\psi_j := \psi^j - \psi^{j-1}$ and $u_j := \mathscr{F}^{-1}[\psi_j\mathscr{F}u]$ for $u \in \mathscr{S}'(G_n)$. We define

$$\mu^\psi_1(u,v) = \sum_{j=2}^{\infty} v_j \sum_{k=0}^{j-2} u_k = \sum_{j=2}^{\infty} v_j u^{j-2},$$

$$\mu^\psi_2(u,v) = \sum_{j=0}^{\infty} u_{j-1}v_j + u_jv_j + u_jv_{j-1},$$

$$\mu^\psi_3(u,v) = \sum_{j=2}^{\infty} u_j \sum_{k=0}^{j-2} v_k = \mu^\psi_1(v,u)$$

for any $u, v \in \mathscr{S}'(G_n)$.

The superscript ψ denotes the dependence on ψ and will often be omitted. The following result allows us to work with the operators μ^ψ_i instead of μ_ψ.

Lemma 3.2.5. *If all μ_l^ψ converge in $\mathscr{S}'(G_n)$, then μ_ψ exists and it holds*

$$\mu_\psi = \mu_1^\psi + \mu_2^\psi + \mu_3^\psi$$

as an identity in $\mathscr{S}'(G_n)$. Furthermore we have

$$\operatorname{supp} \mathscr{F}[v_j u^{j-2}] \subset \{(k,\xi) \in \widehat{G}_n \mid B2^{j-1} \le |k,\xi|_{b,\bar{a}} \le A2^j\} \ and$$
$$\operatorname{supp} \mathscr{F}[u_{j-1}v_j + u_j v_j + u_j v_{j-1}] \subset \{(k,\xi) \in \widehat{G}_n \mid |k,\xi|_{b,\bar{a}} \le A2^j\},$$

for some constants $A, B > 0$ depending on the support of ψ.

Proof: Convergence follows directly from

$$\mu_1^\psi + \mu_2^\psi + \mu_3^\psi = \lim_{m\to\infty} \left(\sum_{j=2}^m v_j u^{j-2} + \sum_{j=0}^m u_{j-1}v_j + u_j v_j + u_j v_{j-1} + \sum_{j=2}^m u_j v^{j-2} \right)$$

$$= \lim_{m\to\infty} \sum_{j=0}^m v_j u^j + u_j v^{j-1} = \lim_{m\to\infty} u^m v^m = \mu_\psi(u,v).$$

Since $\psi \in C_0^\infty(\mathbb{R}^{n+1})$ satisfies the assumptions above equation (3.23) it holds

$$\operatorname{supp} \mathscr{F}u_j \subset \{(k,\xi) \in \widehat{G}_n \mid R_2 2^{j-1} \le |k,\xi|_{b,\bar{a}} \le R_1 2^j\},$$
$$\operatorname{supp} \mathscr{F}u^j \subset \{(k,\xi) \in \widehat{G}_n \mid |k,\xi|_{b,\bar{a}} \le R_1 2^j\},$$

for $j \ge 1$. Note that the convolution of compactly supported distributions is well-defined and the Fourier transform of a product is given by the convolution of the Fourier transform. Because the support of convolutions is contained in the sum of the supports we obtain

$$\mathscr{F}[v_j u^{j-2}] \subset \operatorname{supp} \mathscr{F}v_j + \operatorname{supp} \mathscr{F}u^{j-2} \subset \left\{ (k,\xi) \in \widehat{G}_n \Big| \frac{2^j}{4}(2R_2 - R_1)| \le |k,\xi|_{b,\bar{a}} \le R_1 \left(2^j + \frac{2^j}{4}\right) \right\}$$

by the triangle inequality of $|k,\xi|_{b,\bar{a}}$. Since $2R_2 - R_1 > 0$ by the properties of ψ we derive the first result, the second follows by the same idea. $\square$

Remark 3.2.6. By rearranging the elements in the operators μ_l^ψ one can get rid of the restrictions made in (3.23). For example if one takes μ_1^ψ to sum over the elements $v_j u^{j-3}$, the restriction on R_1 and R_2 becomes $4R_2 > R_1$. But as these restrictions do not cause any difficulties and coincide with the ones made in the definition of Triebel-Lizorkin spaces we will continue with the restrictions made above equation (3.23).

As a first result we will show that μ is given by the standard product $u \cdot v$ for $v \in \mathscr{S}(G_n)$ and $u \in \mathscr{S}'(G_n)$, and hence is an extension of this pairing.

Lemma 3.2.7. *For $v \in \mathscr{S}(G_n)$ and $u \in \mathscr{S}'(G_n)$ it holds $\mu(u,v) = u \cdot v$ as an identity in $\mathscr{S}'(G_n)$.*

Proof: Lemma 1.6.2 yields convergence of $(1 - \psi^m)\mathscr{F}\varphi$ to 0 in $\mathscr{S}(\widehat{G}_n)$ for every $\varphi \in \mathscr{S}(G_n)$ and $m \to \infty$ since ψ satisfies (1.21) by (3.23). Now the continuity properties of the Fourier transform imply that

$$\mathscr{F}_{\widehat{G}_n}^{-1}[(\mathscr{F}^{-1}[\psi^m \mathscr{F}v] - v)\varphi] = \mathscr{F}_{\widehat{G}_n}^{-1}[(v^m - v)\varphi]$$

converges to 0 in $\mathscr{S}(\widehat{G}_n)$ for every $v, \varphi \in \mathscr{S}(G_n)$. With $\varphi, v \in \mathscr{S}(G_n)$ and $u \in \mathscr{S}'(G_n)$ we have

$$\langle v^m u^m, \varphi \rangle - \langle v \cdot u, \varphi \rangle = \langle u^m, v^m \varphi \rangle - \langle \mathscr{F}u, \mathscr{F}_{\widehat{G}_n}^{-1}[\varphi v] \rangle = \langle \mathscr{F}u, \psi^m \mathscr{F}_{\widehat{G}_n}^{-1}[v^m \varphi] - \mathscr{F}_{\widehat{G}_n}^{-1}[\varphi v] \rangle$$
$$= \langle \mathscr{F}u, (\psi^m - 1)\mathscr{F}_{\widehat{G}_n}^{-1}[(v^m - v)\varphi] + (\psi^m - 1)\mathscr{F}_{\widehat{G}_n}^{-1}[v\varphi] + \mathscr{F}_{\widehat{G}_n}^{-1}[(v^m - v)\varphi] \rangle.$$

By the stated convergence the last two arguments converge to 0 in $\mathscr{S}(\widehat{G}_n)$. The term

$$\langle (\psi^m - 1)\mathscr{F}u, \mathscr{F}_{\widehat{G}_n}^{-1}[(v^m - v)\varphi] \rangle \to 0 \quad \text{for } m \to \infty$$

by the uniform boundedness of the sequence of distributions $\{(\psi^m - 1)\mathscr{F}u\}_{m \in \mathbb{N}} \subset \mathscr{S}'(\widehat{G}_n)$ and the convergence of $\mathscr{F}_{\widehat{G}_n}^{-1}[(v^m - v)\varphi]$ to 0. Hence we derive

$$\langle v^m u^m, \varphi \rangle - \langle v \cdot u, \varphi \rangle \to 0 \quad \text{for } m \to \infty.$$

This holds for every ψ from Definition 3.2.1 and because it is independent of ψ we conclude $\mu(u, v) = u \cdot v$. $\qquad\square$

The following result can be found in Johnsen, Munch Hansen and Sickel [53, Lemma 13] in $\mathbb{R}^n$ and we adopt their proof to the setting of G_n.

Theorem 3.2.8. *Let $1 < q, \vec{p} < \infty$, $1 < r \leq \infty$ and $s, s_1 \in \mathbb{R}$ such that $s_1 > |s|$. Then*

(3.25)
$$\|\mu(u, v)\|_{\mathrm{F}_{(q,\vec{p}),r}^{s,(b,\vec{a})}(G_n)} \lesssim \|u\|_{\mathrm{F}_{(q,\vec{p}),r}^{s,(b,\vec{a})}(G_n)} \|v\|_{\mathrm{B}_{\infty,\infty}^{s_1,(b,\vec{a})}(G_n)}$$

holds for all $u \in \mathrm{F}_{(q,\vec{p}),r}^{s,(b,\vec{a})}(G_n)$ and $v \in \mathrm{B}_{\infty,\infty}^{s_1,(b,\vec{a})}(G_n)$.

Proof: We first show estimates of μ_l^{ψ} for $l = 1, 2, 3$ for a fixed ψ. We will carry out the proof with $r < \infty$, but it is easily checked that the arguments can be copied for $r = \infty$. We take ψ equal to ψ_0 from (2.25), in particular it satisfies (3.23). Lemma 3.2.5 allows an application of Lemma 3.1.12 to obtain

$$\|\mu_3^{\psi}(u, v)\|_{\mathrm{F}_{(q,\vec{p}),r}^{s,(b,\vec{a})}(G_n)} \lesssim \left\| \left(\sum_{j=2}^{\infty} 2^{jsr} |u_j v^{j-2}|^r \right)^{\frac{1}{r}} \right\|_{\mathrm{L}_{q,\vec{p}}(G_n)}$$
$$\leq \left\| \left(\sum_{j=2}^{\infty} 2^{jsr} |u_j|^r \right)^{\frac{1}{r}} \right\|_{\mathrm{L}_{q,\vec{p}}(G_n)} \left\| \sup_{j \in \mathbb{N}_0} |v^{j-2}| \right\|_{\mathrm{L}_{\infty}(G_n)}$$
$$\leq \|u\|_{\mathrm{F}_{(q,\vec{p}),r}^{s,(b,\vec{a})}(G_n)} \left\| \sum_{j=0}^{\infty} |v_j| \right\|_{\mathrm{L}_{\infty}(G_n)} \leq \|u\|_{\mathrm{F}_{(q,\vec{p}),r}^{s,(b,\vec{a})}(G_n)} \|v\|_{\mathrm{B}_{\infty,1}^{0,(b,\vec{a})}(G_n)}$$
$$\leq \|u\|_{\mathrm{F}_{(q,\vec{p}),r}^{s,(b,\vec{a})}(G_n)} \|v\|_{\mathrm{B}_{\infty,\infty}^{s_1,(b,\vec{a})}(G_n)},$$

where the last step is an application of Theorem 3.1.8 a) since $s_1 > 0$. Next we define $s_2 := s_1 + s > 0$ and apply Lemma 3.1.12 to derive

$$\|\mu_2^{\psi}(u, v)\|_{\mathrm{F}_{(q,\vec{p}),r}^{s_2,(b,\vec{a})}(G_n)} \lesssim \left\| \left(\sum_{j=0}^{\infty} 2^{js_2 r} |u_{j-1} v_j + u_j v_j + u_j v_{j-1}|^r \right)^{\frac{1}{r}} \right\|_{\mathrm{L}_{q,\vec{p}}(G_n)}$$
$$\lesssim \sup_{j \in \mathbb{N}_0} 2^{js_1} \|v_j\|_{\mathrm{L}_{\infty}(G_n)} \left\| \left(\sum_{j=0}^{\infty} 2^{jsr} |u_j|^r \right)^{\frac{1}{r}} \right\|_{\mathrm{L}_{q,\vec{p}}(G_n)}$$
$$= \|u\|_{\mathrm{F}_{(q,\vec{p}),r}^{s,(b,\vec{a})}(G_n)} \|v\|_{\mathrm{B}_{\infty,\infty}^{s_1,(b,\vec{a})}(G_n)}.$$

For μ_1^ψ we take $t \in (s, s_1)$ such that $t - s_1 < s$ to obtain

$$\|\mu_1^\psi(u,v)\|_{\mathrm{F}_{(q,\vec{p}),r}^{t,(b,\vec{a})}(G_n)} \lesssim \left\|\left(\sum_{j=2}^\infty 2^{jtr}|v_j u^{j-2}|^r\right)^{\frac{1}{r}}\right\|_{\mathrm{L}_{q,\vec{p}}(G_n)}$$

$$\leq \left\|\left(\sum_{j=0}^\infty 2^{j(t-s_1)r}|u^j|^r\right)^{\frac{1}{r}}\right\|_{\mathrm{L}_{q,\vec{p}}(G_n)} \sup_{j\in\mathbb{N}_0} 2^{js_1}\|v_j\|_{\mathrm{L}_\infty(G_n)}.$$

An application of Lemma 5.1.5 and Theorem 3.1.8 to $u^j = \sum_{k=0}^j u_k$ and $t - s_1 < 0$ yields

$$\|\mu_1^\psi(u,v)\|_{\mathrm{F}_{(q,\vec{p}),r}^{t,(b,\vec{a})}(G_n)} \lesssim \left\|\left(\sum_{j=0}^\infty 2^{j(t-s_1)r}|u_j|^r\right)^{\frac{1}{r}}\right\|_{\mathrm{L}_{q,\vec{p}}(G_n)} \sup_{j\in\mathbb{N}_0} 2^{js_1}\|v_j\|_{\mathrm{L}_\infty(G_n)}$$

$$\lesssim \|u\|_{\mathrm{F}_{(q,\vec{p}),r}^{s,(b,\vec{a})}(G_n)}\|v\|_{\mathrm{B}_{\infty,\infty}^{s_1,(b,\vec{a})}(G_n)}.$$

Since $t, s_2 > s$ we apply Theorem 3.1.8 a) to derive

(3.26) $$\|\mu^\psi(u,v)\|_{\mathrm{F}_{(q,\vec{p}),r}^{s,(b,\vec{a})}(G_n)} \lesssim \|u\|_{\mathrm{F}_{(q,\vec{p}),r}^{s,(b,\vec{a})}(G_n)}\|v\|_{\mathrm{B}_{\infty,\infty}^{s_1,(b,\vec{a})}(G_n)}.$$

Hence, we are left to prove independence of the function ψ. We first consider $r < \infty$. Then, by Lemma 3.1.17, the space $\mathscr{S}(G_n)$ is dense in $\mathrm{F}_{(q,\vec{p}),r}^{s,(b,\vec{a})}(G_n)$ and by Lemma 3.2.7 we get $\mu^\psi(u,v) = \mu(u,v) = u \cdot v$ for $u \in \mathscr{S}(G_n)$ and $v \in \mathrm{B}_{\infty,\infty}^{s_1,(b,\vec{a})}(G_n)$. So by (3.26) the operator $\mu(u,v)$ has a unique extension from $\mathscr{S}(G_n)$ to $\mathrm{F}_{(q,\vec{p}),r}^{s,(b,\vec{a})}(G_n)$; this yields independence with respect to the function ψ and hence the result for $r < \infty$. Independence for $r = \infty$ follows by the embedding $\mathrm{F}_{(q,\vec{p}),\infty}^{s,(b,\vec{a})}(G_n) \hookrightarrow \mathrm{F}_{(q,\vec{p}),1}^{s-\varepsilon,(b,\vec{a})}(G_n)$, from Theorem 3.1.8 a), since μ_ψ is independent of ψ in this case. $\square$

In general, the operator μ can differ from the classical product $f \cdot g$, see Johnsen [51] for some examples. But if both elements have enough regularity we have $\mu(f,g) = f \cdot g$, which we will show in the following corollary.

Corollary 3.2.9. *Let $s_1 > s > 0$, $1 < q, \vec{p} < \infty$, and $1 < r \leq \infty$. The operator μ coincides with the classical product $u \cdot v$, i.e., $\mu(u,v) = u \cdot v$, and therefore one has the estimate*

$$\|u \cdot v\|_{\mathrm{F}_{(q,\vec{p}),r}^{s,(b,\vec{a})}(G_n)} \lesssim \|u\|_{\mathrm{F}_{(q,\vec{p}),r}^{s,(b,\vec{a})}(G_n)}\|v\|_{\mathrm{B}_{\infty,\infty}^{s_1,(b,\vec{a})}(G_n)}.$$

Proof: Since $s > 0$ Theorem 3.1.8 a) yields the embedding $\mathrm{F}_{(q,\vec{p}),r}^{s,(b,\vec{a})}(G_n) \hookrightarrow \mathrm{F}_{(q,\vec{p}),2}^{0,(b,\vec{a})}(G_n)$ and Lemma 3.2.2 implies convergence of u^m to u in $\mathrm{F}_{(q,\vec{p}),2}^{0,(b,\vec{a})}(G_n)$. Hence, Proposition 3.1.24 a) implies that u^m converges to u in $\mathrm{L}_{q,\vec{p}}(G_n)$. Because $s_1 > 0$ we derive the embedding $\mathrm{B}_{\infty,\infty}^{s_1,(b,\vec{a})}(G_n) \hookrightarrow \mathrm{B}_{\infty,1}^{0,(b,\vec{a})}(G_n)$ and therefore by Lemma 3.2.3 convergence of v^m to v in $\mathrm{B}_{\infty,1}^{0,(b,\vec{a})}(G_n)$. With the estimate

$$\|v - v^m\|_{\mathrm{L}_\infty(G_n)} \leq \sum_{j=0}^\infty \|\mathscr{F}^{-1}\varphi_j\mathscr{F}(v - v^m)\|_{\mathrm{L}_\infty(G_n)} = \|v - v^m\|_{\mathrm{B}_{\infty,1}^{0,(b,\vec{a})}(G_n)}$$

we conclude convergence in $\mathrm{L}_\infty(G_n)$. Since $u \in \mathrm{L}_{q,\vec{p}}(G_n)$ and $v \in \mathrm{L}_\infty(G_n)$ we have $u \cdot v \in \mathscr{S}'(G_n)$. It holds

$$u^m v^m - uv = (u^m - u)(v^m - v) + (u^m - u)v + u(v^m - v),$$

and hence $u^m v^m$ converges to uv in $L_{q,\vec{p}}$ and therefore in $\mathscr{S}'(G_n)$. This implies $\mu(u,v) = u \cdot v$ because the limit is unique and thus the estimate by Theorem 3.2.8. $\qquad\square$

As we have seen that the space $B^{s,(b,\vec{a})}_{\infty,\infty}(G_n)$ is of importance, we give an additional way to check if $u \in \mathscr{S}'(G_n)$ is an element of that space.

Proposition 3.2.10. *Let $s \in \mathbb{R}$ and $1 \leq r \leq \infty$. A distribution $u \in \mathscr{S}'(G_n)$ is an element of $B^{s,(b,\vec{a})}_{\infty,r}(G_n)$ if and only if $u \circ \pi$ is an element of $B^{s,(b,\vec{a})}_{\infty,r}(\mathbb{R} \times \mathbb{R}^n)$. Furthermore, we have the estimate*

$$\|u\|_{B^{s,(b,\vec{a})}_{\infty,r}(G_n)} \lesssim \|u \circ \pi\|_{B^{s,(b,\vec{a})}_{\infty,r}(\mathbb{R}\times\mathbb{R}^n)} \lesssim \|u\|_{B^{s,(b,\vec{a})}_{\infty,r}(G_n)}.$$

Proof: From Lemma 1.5.12 we obtain the identity

$$\mathscr{F}^{-1}_{\mathbb{R}^{n+1}}\big[\psi_j \mathscr{F}_{\mathbb{R}^{n+1}}(u \circ \pi)\big] = \sum_{l \in \mathbb{Z}} e^{i\frac{2\pi}{T}lt} \otimes \mathscr{F}^{-1}_{\mathbb{R}^n}\Big[\psi_j\Big(\frac{2\pi}{T}l, x\Big)(\mathscr{F}u)_l\Big].$$

Hence if $u \circ \pi \in B^{s,(b,\vec{a})}_{\infty,r}(\mathbb{R} \times \mathbb{R}^n)$ then Corollary 1.5.13 implies that

$$\left(\sum_{j=0}^{\infty} 2^{jsr}\big\|\mathscr{F}^{-1}[\tilde{\varphi}_j\mathscr{F}u]\big\|^r_{L_\infty(G_n)}\right)^{\frac{1}{r}} = \|u \circ \pi\|_{B^{s,(b,\vec{a})}_{\infty,r}(\mathbb{R}\times\mathbb{R}^n)} < \infty$$

with $\tilde{\varphi}_j := \psi_j\big|_{\frac{2\pi}{T}\mathbb{Z}\times\mathbb{R}^n}$. Therefore, Lemma 3.1.15 yields

$$\|u\|_{B^{s,(b,\vec{a})}_{\infty,r}(G_n)} \lesssim \|u \circ \pi\|_{B^{s,(b,\vec{a})}_{\infty,r}(\mathbb{R}\times\mathbb{R}^n)}.$$

The converse estimate follows by the same arguments since Lemma 3.1.15 stays valid by Corollary 3.1.7. $\qquad\square$

3.3 Extension Operators

As stated in the introduction of the chapter, we are going to construct a right inverse to the trace operator with respect to the variable x_n, *i.e.*, for a distribution $v \in \mathscr{S}'(G_{n-1})$ we want to find a distribution $u \in \mathscr{S}'(G_n)$ such that $u\big|_{x_n=0}$ is well-defined and is equal to v. For the construction we take $\tilde{\varphi}_j(k, \xi') := \varphi_j(k, \xi', 0)$, which satisfies $\sum_j \tilde{\varphi}_j = 1$ on $\widehat{G}_{n-1}$ and

$$\operatorname{supp} \tilde{\varphi}_j \subset \left\{ (k, \xi') \in \widehat{G}_{n-1} \big| \ 2^{j-1} \leq |(k, \xi')|_{b, \vec{a}'} \leq \frac{3}{2} \cdot 2^j \right\}.$$

For $v \in \mathscr{S}'(G_{n-1})$ we define $v_j = \mathscr{F}^{-1} \tilde{\varphi}_j \mathscr{F} v$ and obtain the identity $v = \sum_{j=0}^{\infty} v_j$ in $\mathscr{S}'(G_{n-1})$ by Lemma 1.6.2. Let $\psi \in \mathscr{S}(\mathbb{R})$ such that $\psi(0) = 1$ and $\operatorname{supp} \mathscr{F}_{\mathbb{R}} \psi \subset [1, 2]$; this can be achieved by taking the inverse Fourier transform of a positive $C_0^{\infty}(\mathbb{R})$ function and normalising its integral to 1. With this we define the extension operator E_n by

$$(3.27) \qquad\qquad E_n v := \sum_{j=0}^{\infty} \psi(2^{j a_n} x_n) v_j(t, x').$$

Throughout this section we assume $s \in \mathbb{R}$, $1 < q, \vec{p} < \infty$, and $1 < r \leq \infty$. We start by showing that E_n is a continuous function with respect to x_n, the trace exists and equals u.

Lemma 3.3.1. *For every $v \in \mathscr{S}'(G_{n-1})$ it holds $E_n v \in \mathscr{S}'(G_n)$ and $\Upsilon : x_n \mapsto E_n v(\cdot, x_n)$ is an element of $C(\mathbb{R}, \mathscr{S}'(G_{n-1}))$. Furthermore, $E_n v\big|_{x_n=0}$ is well-defined and coincides with v.*

Proof: We want to apply Lemma 1.6.1. Hence, we need to estimate the elements of the sum. By Lemma 1.5.10 and using the properties of the support of $\mathscr{F}_{G_{n-1}} v_j$ we derive

$$v_j(t, x') = \sum_{|m| \leq \frac{3b}{2b} 2^{jb}} \mathscr{F}_{\mathbb{R}^{n-1}}^{-1}[(\mathscr{F}_{G_{n-1}} v_j)_m] \otimes e^{i \frac{2\pi}{T} mt}.$$

Because $(\mathscr{F}_{G_{n-1}} v_j)_m \in \mathcal{E}'(\mathbb{R}^{n-1})$, the space of compactly supported distributions, Hörmander [46, Theorem 7.1.14] yields the identity

$$\mathscr{F}_{\mathbb{R}^{n-1}}^{-1}[(\mathscr{F}_{G_{n-1}} v_j)_m](x') = c_n \langle (\mathscr{F}_{G_{n-1}} v_j)_m, e^{i x' \cdot \xi'} \rangle = c_n \langle \mathscr{F} v, \tilde{\varphi}_j \delta_m e^{i x' \cdot \xi'} \rangle.$$

By applying Lemma 1.5.3 to this identity we obtain an $l \in \mathbb{N}$ such that

$$|v_j(t, x')| \lesssim \sum_{|m| \leq \frac{3b}{2b} 2^{jb}} |\langle \mathscr{F} v, \tilde{\varphi}_j \delta_m e^{i x' \cdot \xi'} \rangle| \lesssim \sum_{|m| \leq \frac{3b}{2b} 2^{jb}} \sup_{(k, \xi') \in \widehat{G}_{n-1}} (1 + |k|)^l (1 + |\xi'|)^l |\hat{D}^l(\tilde{\varphi}_j \delta_m e^{i x' \cdot \xi'})|.$$

We can estimate $|k|$ by $3^b 2^{jb-b}$ as otherwise the functions $\delta_m(\cdot)$ are zero, afterwards estimate the delta distribution by 1 to get rid of the sum. This yields

$$|v_j(t, x')| \lesssim \sup_{(k, \xi) \in \widehat{G}_{n-1}} 2^{jb(l+1)} (1 + |\xi|)^l |D^l(\tilde{\varphi}_j e^{i x' \cdot \xi'})|.$$

Next, we estimate all derivatives up to order l. For arbitrary $\alpha \in \mathbb{N}_0^{n-1}$ with $|\alpha| \leq l$ we have

$$|D_{\xi'}^{\alpha}(\tilde{\varphi}_j e^{i x' \cdot \xi'})| \lesssim \sum_{\gamma \leq \alpha} |D^{\gamma}[\tilde{\varphi}_j](i x')^{\alpha - \gamma} e^{i x' \cdot \xi'}| \lesssim (1 + |x'|)^l$$

by (3.2). The results of Proposition 1.4.2 and $(k, \xi) \in \text{supp } \tilde{\varphi}_j$ yield the estimate

$$1 + |\xi'| \leq 1 + \sum_{i=1}^{n-1} |\xi_i| \leq 1 + \sum_{i=1}^{n-1} |(k, \xi)|_{b, \bar{a}}^{a_i} \leq 1 + \sum_{i=1}^{n-1} \left(\frac{3}{2}\right)^{a_i} 2^{j a_i} \leq 2^{j|\bar{a}|} + 2^{j|\bar{a}|} \sum_{i=1}^{n-1} \left(\frac{3}{2}\right)^{a_i} \lesssim 2^{j|\bar{a}|}.$$

By combining the previous results we derive

$$|v_j(t, x')| \lesssim 2^{j 2 b l} 2^{j|\bar{a}| l} (1 + |x'|)^l \leq 2^{jm} (1 + |x'|)^m,$$

for some $m \in \mathbb{N}$. Since $\psi \in \mathscr{S}(\mathbb{R})$ we conclude

$$|\psi(2^{j a_n} x_n) v_j(t, x')| \lesssim 2^{jm} (1 + |x'|)^m$$

and hence the estimate in the assumptions of Lemma 1.6.1 is satisfied. To prove the support condition we note that $\mathscr{F}_{\mathbb{R}}[\psi(2^{j a_n} \cdot)] = 2^{-j a_n} \mathscr{F}_{\mathbb{R}}[\psi](2^{-j a_n} \cdot)$, and hence it holds

$$\text{supp } \mathscr{F}_{G_n}[\psi(2^{j a_n} \cdot) v_j] = \text{supp } \mathscr{F}_{\mathbb{R}}[\psi(2^{j a_n} \cdot)] \mathscr{F}_{G_{n-1}}[v_j]$$

$$\subset \{\xi_n \in \mathbb{R} \mid 2^{j a_n} \leq |\xi_n| \leq 2 \cdot 2^{j a_n}\} \times \left\{(k, \xi') \in \widehat{G}_{n-1} \mid 2^{j-1} \leq |(k, \xi')|_{b, \bar{a}'} \leq \frac{3}{2} \cdot 2^j\right\}$$

$$(3.28) \qquad \subset \left\{(k, \xi) \in \widehat{G}_n \mid 2 \cdot 2^{j-1} \leq |(k, \xi)|_{b, \bar{a}} \leq (\frac{3}{2} + 2^{\frac{1}{a_n}}) 2^j\right\},$$

where the last inclusion follows from Proposition 1.4.2 c) and d) since

$$2^j \leq |\xi_n|^{\frac{1}{a_n}} \leq |(k, \xi)|_{b, \bar{a}} \leq |(k, \xi', 0)|_{b, \bar{a}} + |(0, \xi_n)|_{b, \bar{a}} = |(k, \xi')|_{b, \bar{a}'} + |\xi_n|^{\frac{1}{a_n}} \leq (\frac{3}{2} + 2^{\frac{1}{a_n}}) 2^j.$$

So we obtain convergence in $\mathscr{S}'(G_n)$ by Lemma 1.6.1.

To show continuity with respect to x_n we note that from equation (1.20) we derive the estimate $|\langle v_j, \lambda \rangle| \lesssim 2^{-jm}$ for $m > 0$ and $\lambda \in \mathscr{S}(G_{n-1})$. Since $\psi(2^{j a_n} x_n)$, $j \in \mathbb{N}_0$, is a continuous and uniformly bounded family of functions, the series $\sum_{j=0}^{\infty} \psi(2^{j a_n} x_n) \langle v_j, \varphi \rangle$ converges uniformly to a bounded and continuous function for every $\varphi \in \mathscr{S}(G_{n-1})$. Convergence in $C(\mathbb{R}, \mathscr{S}'(G_{n-1}))$ yields convergence in $\mathscr{S}'(G_n)$ and hence Υ is an element of $C(\mathbb{R}, \mathscr{S}'(G_{n-1}))$. This implies that the value at $x_n = 0$ of $E_n v$ is well-defined, and with $\psi(0) = 1$ we obtain

$$(3.29) \qquad \langle [E_n v](t, x) \big|_{x_n = 0}, \lambda \rangle = \langle [E_n v](t, x', 0), \lambda \rangle = \langle \sum_{j=0}^{\infty} v_j(t, x'), \lambda \rangle = \langle v, \lambda \rangle$$

by the properties of $\tilde{\varphi}_j$ for every $\lambda \in \mathscr{S}(G_{n-1})$. $\qquad \square$

We now state one of the main theorems of this section. Since we allow the trace to be a distribution we can define an extension operator for every $s \in \mathbb{R}$.

Theorem 3.3.2. *The extension operator E_n defined by equation (3.27) is a continuous map from* $\mathrm{F}_{(q, \vec{p}'), p_n}^{s - \frac{a_n}{p_n}, (b, \vec{a}')}(G_{n-1})$ *to* $\mathrm{F}_{(q, \vec{p}), r}^{s, (b, \vec{a})}(G_n)$ *and satisfies the estimate*

$$(3.30) \qquad \|E_n v\|_{\mathrm{F}_{(q, \vec{p}), r}^{s, (b, \vec{a})}(G_n)} \lesssim \|v\|_{\mathrm{F}_{(q, \vec{p}'), p_n}^{s - \frac{a_n}{p_n}, (b, \vec{a}')}(G_{n-1})}.$$

Proof: By Lemma 3.1.13 it suffices to show the estimate

$$(3.31) \qquad \left\| \{2^{js}\psi(2^{ja_n}\cdot)v_j\}_{j\in\mathbb{N}_0} \right\|_{L_{q,\vec{p}}(G_n;\ell_r)} \leq c\|v\|_{\mathrm{F}^{s-\frac{a_n}{p_n},(b,\vec{a}')}_{(q,\vec{p}'),p_n}(G_{n-1})},$$

since the support of the Fourier transform of each summand satisfies the needed properties, see (3.28). By the monotonicity of the ℓ_r-spaces it suffices to show the estimate for $r < p_n$. Since $\psi \in \mathscr{S}(\mathbb{R})$ we conclude the estimate $|\psi(2^{ja_n}x_n)| \leq \rho_{0,0,1}(\psi)|2^{ja_n}x_n|^{-1}$ for every $x_n \neq 0$. Because $p_n > 1$ we derive

$$\int\limits_{|x_n|>1} \left(\sum_{j=0}^{\infty} |2^{sj}\psi(2^{ja_n}x_n)v_j(t,x')|^r \right)^{\frac{p_n}{r}} dx_n \lesssim \int\limits_{|x_n|>1} \left(\sum_{j=0}^{\infty} |2^{sj}v_j(t,x')|^r 2^{-ja_nr} \right)^{\frac{p_n}{r}} |x_n|^{-p_n} dx_n$$

$$\lesssim \left(\sup_{j\in\mathbb{N}_0} 2^{\left(s-\frac{a_n}{p_n}\right)j} |v_j(t,x')| \right)^{p_n} \left(\sum_{j=0}^{\infty} 2^{ja_nr\left(\frac{1}{p_n}-1\right)} \right)^{\frac{p_n}{r}} \lesssim \left(\sup_{j\in\mathbb{N}_0} 2^{j\left(s-\frac{a_n}{p_n}\right)} |v_j(t,x')| \right)^{p_n}$$

$$\leq \sum_{j=0}^{\infty} |2^{j\left(s-\frac{a_n}{p_n}\right)}v_j(t,x')|^{p_n}.$$

For the remaining integral we define $A_k := \{x_n \in \mathbb{R} \mid 2^{-(k+1)a_n} \leq |x_n| \leq 2^{-ka_n}\}$ to obtain

$$\int\limits_{|x_n|\leq 1} \left(\sum_{j=0}^{\infty} |2^{sj}\psi(2^{ja_n}x_n)v_j(t,x')|^r \right)^{\frac{p_n}{r}} dx_n = \sum_{k=0}^{\infty} \int\limits_{A_k} \left(\sum_{j=0}^{\infty} |2^{sj}\psi(2^{ja_n}x_n)v_j(t,x')|^r \right)^{\frac{p_n}{r}} dx_n$$

$$\leq (2-2^{1-a_n}) \sum_{k=0}^{\infty} 2^{-ka_n} \left(\sum_{j=0}^{k} |2^{sj}v_j(t,x')|^r \|\psi\|_\infty^r + \sum_{j=k+1}^{\infty} c(\psi)^r |2^{j(s-a_n)+(k+1)a_n}v_j(t,x')|^r \right)^{\frac{p_n}{r}}$$

$$\lesssim \sum_{k=0}^{\infty} 2^{-kp_n\frac{a_n}{p_n}} \left(\sum_{j=0}^{k} |2^{sj}v_j(t,x')|^r \right)^{\frac{p_n}{r}} + \sum_{k=0}^{\infty} 2^{kp_n\left(a_n-\frac{a_n}{p_n}\right)} \left(\sum_{j=k+1}^{\infty} |2^{j(s-a_n)}v_j(t,x')|^r \right)^{\frac{p_n}{r}}.$$

To the last line we apply Lemma 5.1.5 with $q = p_n$, which is possible as $\frac{a_n}{p_n}$ and $a_n - \frac{a_n}{p_n}$ are both positive. Hence it holds

$$\int_{|x_n|\leq 1} \left(\sum_{j=0}^{\infty} |2^{sj}\psi(2^{ja_n}x_n)v_j(t,x')|^r \right)^{\frac{p_n}{r}} dx_n \lesssim \sum_{j=0}^{\infty} |2^{j\left(s-\frac{a_n}{p_n}\right)}v_j(t,x')|^{p_n}.$$

Combining both estimates we conclude

$$\int_{\mathbb{R}} \left(\sum_{j=0}^{\infty} |2^{sj}\psi(2^{ja_n}x_n)v_j(t,x')|^r \right)^{\frac{p_n}{r}} dx_n \lesssim \sum_{j=0}^{\infty} |2^{j\left(s-\frac{a_n}{p_n}\right)}v_j(t,x')|^{p_n}.$$

Applying the p_n-th root and the $L_{q,\vec{p}}(G_{n-1})$-norm yield the estimate from equation (3.31) and hence the result. □

As a next step we construct an extension operator to higher order derivative terms, *i.e.* for a distribution $v \in \mathscr{S}'(G_{n-1})$ we want to find $u \in \mathscr{S}'(G_n)$ such that $\partial_n^m u|_{x_n=0}$ is well-defined and equal to v for $m \in \mathbb{N}$. The idea is similar to the one used in the construction of E_n: We take a function $\psi_m \in \mathscr{S}(\mathbb{R})$ such that $\partial^m \psi_m(0) = 1$ and $\operatorname{supp} \mathscr{F} \psi_m \subset [1,2]$ and define

$$(3.32) \qquad E_n^m v := \sum_{j=0}^{\infty} 2^{-j a_n m} \psi_m(2^{j a_n} x_n) v_j(t, x').$$

The functions v_j are constructed in the same way as before for a distribution $v \in \mathscr{S}'(G_{n-1})$. This operator works similar to E_n. We state all needed properties in the next lemma.

Lemma 3.3.3. *The operator E_n^m is well-defined and for every $v \in \mathscr{S}'(G_n)$ the function $\Upsilon_m :$ $x_n \mapsto E_n^m v$ is an element of $\mathrm{C}^m(\mathbb{R}, \mathscr{S}'(G_{n-1}))$. Additionally, $\partial_n^m E_n^m v\big|_{x_n=0}$ is well-defined and equal to v and the estimate*

$$(3.33) \qquad \|E_n^m v\|_{\mathrm{F}^{s,(b,\vec{a})}_{(q,\vec{p}),r}(G_n)} \leq c \|v\|_{\mathrm{F}^{s-a_n m - \frac{a_n}{p_n},(b,\vec{a}')}_{(q,\vec{p}'),p_n}(G_{n-1})}$$

holds.

Proof: Convergence follows directly from Lemma 3.3.1 as the additional term $2^{-j a_n m}$ only helps with convergence. By the same lemma we derive that $\sum_{j=0}^{\infty} 2^{-j a_n m} \partial_n^l[\psi_m(2^{j a_n} x_n)] v_j(t, x')$ converges uniformly to a continuous function for every $l \in \mathbb{N}$ satisfying $l \leq m$. This yields

$$\partial_n^l E_n^m = \sum_{j=0}^{\infty} 2^{-j a_n (m-l)} \psi_m(2^{j a_n} x_n) v_j(t, x')$$

and hence $\Upsilon_m \in \mathrm{C}^m(\mathbb{R}, \mathscr{S}'(G_n))$. By the same arguments as before we obtain the same regularity for E_n^m, and thus the value $\partial_n^m E_n v$ is well-defined and it holds

$$\big\langle \partial_n^m E_n v\big|_{x_n=0}, \varphi \big\rangle = \big\langle \sum_{j=0}^{\infty} \partial_n^m \psi_m(0) v_j(t,x'), \varphi \big\rangle = \big\langle \sum_{j=0}^{\infty} v_j(t,x'), \varphi \big\rangle = \langle v, \varphi \rangle$$

by the properties of ψ_m and v_j for every $\varphi \in \mathscr{S}(G_{n-1})$. The estimate from (3.33) follows word by word from the proof of Theorem 3.3.2 as the extra factor $2^{-j a_n m}$ only yields regularity and causes no need for modification in the proof. $\qquad \square$

Remark 3.3.4. One of the problems stated in the introduction of this chapter was the question of whether the extension preserves pure periodicity. By the construction of E_n respectively E_n^m this reduces to $\delta_0 \mathscr{F}_{\mathbb{T}}[\psi(2^{j a_n} \cdot) v_j] = 0$ if $\delta_0 \mathscr{F}_{\mathbb{T}} v = 0$ for every $j \in \mathbb{N}_0$, because the convergence of the series trivially preserves this property. For $\lambda \in \mathscr{S}(\widehat{G}_n)$ it holds

$$\big\langle \delta_0 \mathscr{F}_{\mathbb{T}}[\psi(2^{j a_n} \cdot) v_j], \lambda \big\rangle = \big\langle \psi(2^{j a_n} x_n), \langle \mathscr{F}^{-1}_{\mathbb{R}^{n-1}}[\tilde{\varphi}_j \mathscr{F}_{G_{n-1}} v], \delta_0 \lambda(\cdot, x_n) \rangle \big\rangle$$
$$= \big\langle \psi(2^{j a_n} x_n), \langle \mathscr{F}_{G_{n-1}} v, \tilde{\varphi}_j \delta_0 \mathscr{F}^{-1}_{\mathbb{R}^{n-1}}[\lambda(\cdot, x_n)] \rangle \big\rangle$$
$$= \big\langle \psi(2^{j a_n} x_n), \langle \delta_0 \mathscr{F}_{\mathbb{T}} v, \mathscr{F}_{\mathbb{R}^{n-1}}[\tilde{\varphi}_j \mathscr{F}^{-1}_{\mathbb{R}^{n-1}}[\lambda(\cdot, x_n)]] \rangle \big\rangle = 0$$

for any $v \in \mathcal{P}_\perp \mathscr{S}(G_n)$. Hence the extension operators E_n and E_n^m preserve pure periodicity.

3.3.1 Extension Operators in Hölder Spaces

The extension of functions in Hölder spaces has been studied for example in Gilbarg and Trudinger [40, Section 6.9]. The approach used there and in other works relies on a partition of unity resulting in estimates that always use the full inhomogeneous norm. The aim of this section is to construct an extension in the fixed setting of a ball $B_r = B_r(0) \subset \mathbb{R}^n$ with radius r such that we get homogeneous estimates and are able to see how the radius r influences the occurring constants. Before we can come to the main results we need to prove some auxiliary results.

Lemma 3.3.5. *Let $h : B_r^c \to \mathbb{R}$ be either $\frac{r^m}{|x|^m}$ or $\frac{-mr^m x_k x_j}{|x|^{m+2}}$. Then*

$$\|h\|_{\widehat{C}^\alpha(B_r^c)} \lesssim r^{-\alpha}$$

for all $j, k \in \{1, 2, \ldots, n\}$, all $r \in \mathbb{R}_+$ and $0 < \alpha \leq 1$. For the function $\varphi_m(x) : B_r^c \to \mathbb{R}^n$ with $\varphi_m(x) = \frac{r^m}{|x|^m} \cdot x$ we have the estimate

$$\|\varphi_m\|_{\widehat{C}^1(B_r^c)} \leq (m+1)$$

for all $r \in \mathbb{R}$.

Proof: We first note that $|h| \leq m$ holds independent of r. With this we get

$$\sup_{\substack{x,y \in B_r^c \\ |x-y| \geq r}} \frac{|h(x) - h(y)|}{|x-y|^\alpha} \leq 2mr^{-\alpha}.$$

Next up we consider the case of $|x - y| \leq r$. We want to use the mean value theorem, but as B_r^c is not convex the proof is not straightforward. As we only consider the case of $|x-y| \leq r$ we need to determine the minimal distance of the connecting line between x and y to the origin. Hence we are dealing with the three points x, y and the origin, thus it suffices to consider the two dimensional case. Here, a simple geometric consideration shows that the distance is at least $\frac{\sqrt{3}}{2} r$. Since ∇h is either $\frac{-mr^m}{|x|^{m+2}} \cdot x$ or $\frac{-(m+2)mr^m x_k x_j}{|x|^{m+4}} \cdot x + \frac{mr^m x_k}{|x|^{m+2}} \cdot e_j + \frac{mr^m x_j}{|x|^{m+2}} \cdot e_k$ we obtain the estimate

$$\frac{|h(x) - h(y)|}{|x-y|^\alpha} \leq \sup_{|\xi| \geq \frac{\sqrt{3}}{2} r} |\nabla h(\xi)| |x-y|^{1-\alpha} \lesssim r^{-\alpha}.$$

Combining these two estimates yields the results for h. Regarding φ_m it holds

$$|\varphi_m(x) - \varphi_m(y)| = r^m \left| \frac{x}{|x|^m} - \frac{y}{|y|^m} \right| \leq \frac{r^m}{|x|^m} |x-y| + \frac{r^m |y|}{|x|^m |y|^m} \left| |y|^m - |x|^m \right|$$

$$\leq |x-y| + \frac{r^m |y| |x-y|}{|x|^m |y|^m} \sum_{k=0}^{m-1} |y|^{m-1-k} |x|^k$$

$$\leq |x-y| + r^m |x-y| \sum_{k=0}^{m-1} r^{-m} = (m+1)|x-y|.$$

This implies the results for φ_m. $\qquad\qquad\square$

Lemma 3.3.6. *The functions $f_j, g : B_r^c \to \mathbb{R}$ defined by $f_j(x) = \frac{x_j}{|x|}$ and $g(x) = \frac{1}{|x|}$ are elements of $\mathrm{C}^{0,\alpha}(B_r^c)$ for all $\alpha \in (0, 1]$ and all $j = 0, 1, \ldots, n$. Furthermore we have*

$$\|f_j\|_{\mathrm{C}^{0,\alpha}(B_r^c)} \leq 3\|f_j\|_{\mathrm{C}^{0,1}(B_r^c)} \leq 3 + \frac{6}{r},$$

$$\|g\|_{\mathrm{C}^{0,\alpha}(B_r^c)} \leq 3\|g\|_{\mathrm{C}^{0,1}(B_r^c)} \leq \frac{3}{r} + \frac{3}{r^2}.$$

Proof: It is clear that $|f_j(x)| \leq 1$ and $|g(x)| \leq r^{-1}$ hold for all $x \in B_r^c$. Furthermore we have

$$\left| \frac{x_j}{|x|} - \frac{y_j}{|y|} \right| = \left| \frac{x_j|y| - y_j|x|}{|x||y|} \right| \leq \left| \frac{|y| - |x|}{|y|} \right| + \left| \frac{|x_j - y_j|}{|y|} \right| \leq \frac{2}{|y|}|x - y| \leq \frac{2}{r}|x - y|,$$

$$\left| \frac{1}{|x|} - \frac{1}{|y|} \right| = \left| \frac{|x| - |y|}{|x||y|} \right| \leq \frac{|x - y|}{|x||y|} \leq \frac{|x - y|}{r^2}.$$

This proves the estimates of the $\mathrm{C}^{0,1}$-norm. For the $\mathrm{C}^{0,\alpha}$-norm we note that

$$\frac{|h(x) - h(y)|}{|x - y|^\alpha} \leq \frac{|h(x) - h(y)|}{|x - y|}$$

for $|x - y| \leq 1$ and

$$\frac{|h(x) - h(y)|}{|x - y|^\alpha} \leq 2\|h\|_{\mathrm{C}^0}$$

for $|x - y| \geq 1$. This implies $\|h\|_{\widehat{\mathrm{C}}^\alpha} \leq \|h\|_{\widehat{\mathrm{C}}^1} + 2\|h\|_{\mathrm{C}^0}$ and hence the result follows. $\qquad\square$

With the preparations out of the way, we can state the main results of this section.

Theorem 3.3.7. *Let $u : B_r \to \mathbb{R}$ be a function in $\mathrm{C}^{2,\alpha}(B_r)$ for $\alpha \in (0, 1]$. We define the function*

$$v(x) = \begin{cases} u(x), & |x| \leq r, \\ 6u\left(\frac{r^2}{|x|^2} \cdot x\right) - 8u\left(\frac{r^3}{|x|^3} \cdot x\right) + 3u\left(\frac{r^4}{|x|^4} \cdot x\right), & |x| > r. \end{cases}$$

Then $v \in \mathrm{C}^{2,\alpha}(\mathbb{R}^n)$ and there exist constants $C_1, C_2(r), C_3 > 0$ such that

$$\|v\|_{\mathrm{C}^l(\mathbb{R}^n)} \leq C_1\|u\|_{\mathrm{C}^l(B_r)},$$

$$\|v\|_{\mathrm{C}^{2,\alpha}(\mathbb{R}^n)} \leq C_2(r)\|u\|_{\mathrm{C}^{2,\alpha}(B_r)},$$

$$\|\nabla v\|_{\widehat{\mathrm{C}}^\alpha(\mathbb{R}^n)} \leq C_3\left(\|\nabla u\|_{\mathrm{C}^0(B_r)}r^{-\alpha} + \|\nabla u\|_{\widehat{\mathrm{C}}^\alpha(B_r)}\right),$$

holds for $l = 0, 1$ and the constants C_1, C_3 are independent of r.

Proof: First, note that u and all the derivatives of u of order up to two are uniformly continuous and therefore the limit at the boundary exists. We define the functions $\varphi_m(x) := \frac{r^m}{|x|^m}x =: \psi_m(x) \cdot x$, and it holds $\psi_m(x) = 1$ for all $x \in \partial B_r$ and therefore $\varphi_m(x) = x$ for all $x \in \partial B_r$. This yields

One of the advantages of working in Triebel-Lizorkin spaces is that there is an easy way to define a trace. By fixing the notation $u_j = \mathscr{F}^{-1}\varphi_j\mathscr{F}u$ for $u \in \mathscr{S}'(G_n)$ we derive from Theorem 1.5.11 that each $u_j \in C^\infty(G_n)$ and hence the value $u_j\big|_{x_n=0}$ is well-defined. So the natural idea to define a trace $T_k u$ is

$$(3.34) \qquad T_k u = \sum_{j=0}^{\infty} u_j\big|_{x_k=0}.$$

We are going to show that the series from equation (3.34) converges in a reasonable sense and coincides with the value $u\big|_{x_k=0}$ where it exists. We restrict the considerations to the case of $k = n$, since this is the important case for the half space, and generally suffices when extending the theory to domains. As in Section 3.3 we consider the case of $s \in \mathbb{R}$, $1 < q, \vec{p} < \infty$, and $1 < r \le \infty$ and start by giving the main estimate of this section.

Proposition 3.4.1. *For $u \in F_{(q,\vec{p}),r}^{s,(b,\vec{a})}(G_n)$ we have the estimate*

$$\sup_{z \in \mathbb{R}} \left\| \left(\sum_{j=0}^{\infty} |2^{(s-\frac{a_n}{p_n})j} u_j(\cdot, z)|^{p_n} \right)^{\frac{1}{p_n}} \right\|_{L_{q,\vec{p}'}(G_{n-1})} \lesssim \|u\|_{F_{(q,\vec{p}),r}^{s,(b,\vec{a})}(G_n)}.$$

Proof: It holds

$$|u_j(t, x', z)| = \frac{(1 + |(d2^j)^{\vec{a}}y|)^n}{(1 + |(d2^j)^{\vec{a}}y|)^n}|u_j(t, x', x_n - y_n)|\Big|_{(y',y_n)=(0,x_n-z)}.$$

By restricting x_n to $[z + 2^{-ja_n}, z + 2^{(1-j)a_n}]$ we derive $x_n - z \in [2^{-ja_n}, 2^{(1-j)a_n}]$, and hence the estimate $(1 + |(d2^j)^{\vec{a}}y|)^n \le (1 + d^{a_n}2^{a_n})^n$ for $y = (y', y_n) = (0, x_n - z)$. By estimating the right-hand side by the maximal function from (1.26) we obtain for $x_n \in [z + 2^{-ja_n}, z + 2^{(1-j)a_n}]$ the estimate

$$|u_j(t, x', z)| \lesssim u_j^*((d2^j)^{\vec{a}}; t, x', x_n).$$

Integrating over these x_n yields

$$(2^{a_n} - 1)2^{-ja_n}|u_j(t, x', z)|^{p_n} = \int_{z+2^{-ja_n}}^{z+2^{(1-j)a_n}} |u_j(t, x', z)|^{p_n} \, dx_n \lesssim \int_{z+2^{-ja_n}}^{z+2^{(1-j)a_n}} |u_j^*((d2^j)^{\vec{a}}; t, x', x_n)|^{p_n} \, dx_n.$$

We multiply by 2^{sjp_n} and sum with respect to j to obtain for any $N, M \in \mathbb{N}$ with $N < M$

$$\sum_{j=N}^{M} 2^{jp_n(s-\frac{a_n}{p_n})}|u_j(t, x', z)|^{p_n} \lesssim \sum_{j=N}^{M} \int_{z+2^{-ja_n}}^{z+2^{(1-j)a_n}} |2^{sj}u_j^*((d2^j)^{\vec{a}}; t, x', x_n)|^{p_n} \, dx_n$$

$$\lesssim \int_{\mathbb{R}} \chi_{\{x_n \in [z, z+2^{(1-N)a_n}]\}} \left(\sup_{k \in \mathbb{N}} |2^{sk}u_k^*((d2^k)^{\vec{a}}; t, x', x_n)| \right)^{p_n} \, dx_n.$$

We note that the term $\chi_{\{x_n \in [z, z+2^{(1-N)a_n}]\}}$ can be omitted, but will be crucial in the proof of Lemma 3.4.2 below, see (3.37). Taking the $\frac{1}{p_n}$-th power on both sides, applying the $L_{q,\vec{p}'}(G_{n-1})$-norm and using the estimate of Proposition 1.7.6 we conclude

$$\left\| \left(\sum_{j=N}^{M} |2^{(s-\frac{a_n}{p_n})j}u_j(\cdot, z)|^{p_n} \right)^{\frac{1}{p_n}} \right\|_{L_{q,\vec{p}'}(G_{n-1})} \lesssim \left\| \chi_{\{x_n \in [z, z+2^{(1-N)a_n}]\}}2^{sj}u_j^*((d2^j)^{\vec{a}}; \cdot) \right\|_{L_{q,\vec{p}}(G_n;\ell_\infty)}$$

$$(3.35) \qquad\qquad\qquad\qquad\qquad \le \left\| 2^{sj}u_j \right\|_{L_{q,\vec{p}}(G_n;\ell_\infty)} \le \|u\|_{F_{(q,\vec{p}),r}^{s,(b,\vec{a})}(G_n)}.$$

Since the right hand side is independent of N, M and z the result follows by monotone convergence and taking the supremum in z. $\qquad\square$

With the previous estimate we can show that the defining series for the operator T_n converges in $L_{q,\vec{p}'}(G_{n-1})$ for every $z \in \mathbb{R}$.

Lemma 3.4.2. *Let $s > \frac{a_n}{p_n}$. For all $u \in F^{s,(b,\vec{a})}_{(q,\vec{p}),r}(G_n)$ we have the estimate*

$$\sup_{z \in \mathbb{R}} \Big\| \sum_{j=0}^{\infty} u_j(\cdot, z) \Big\|_{L_{q,\vec{p}'}(G_{n-1})} \lesssim \|u\|_{F^{s,(b,\vec{a})}_{(q,\vec{p}),r}(G_n)}.$$

Consequently the series on the left-hand side converges in $L_{q,\vec{p}'}(G_{n-1})$ for every $z \in \mathbb{R}$.

Proof: First, we show that $\Big\{ \sum_{j=0}^{k} u_j(t, x', z) \Big\}_{k \in \mathbb{N}}$ is a Cauchy sequence in $L_{q,\vec{p}'}(G_{n-1})$ for every $z \in \mathbb{R}$. For $N, M \in \mathbb{N}$ with $N \le M$ it holds

$$\Big\| \sum_{j=N}^{M} u_j(\cdot, z) \Big\|_{L_{q,\vec{p}'}(G_{n-1})} \le \Big\| \Big(\sum_{j=N}^{M} |2^{(s-\frac{a_n}{p_n})j} u_j(\cdot, z)|^{p_n} \Big)^{\frac{1}{p_n}} \Big\| 2^{-(s-\frac{a_n}{p_n})(\cdot)} \Big\|_{\ell_{p_n'}} \Big\|_{L_{q,\vec{p}'}(G_{n-1})}$$

$$(3.36) \qquad\qquad \lesssim \Big\| \Big(\sum_{j=N}^{M} |2^{(s-\frac{a_n}{p_n})j} u_j(\cdot, z)|^{p_n} \Big)^{\frac{1}{p_n}} \Big\|_{L_{q,\vec{p}'}(G_{n-1})}$$

An application of the first part of (3.35) yields

$$(3.37) \qquad \Big\| \sum_{j=N}^{M} u_j(\cdot, z) \Big\|_{L_{q,\vec{p}'}(G_{n-1})} \lesssim \Big\| \chi_{\{x_n \in [z, z+2^{(1-N)a_n}]\}} 2^{sj} u_j^*((d2^j)^{\vec{a}}; \cdot) \Big\|_{L_{q,\vec{p}}(G_n;\ell_\infty)}.$$

Additionally, by the second part of (3.35), we know that $\sup_{j \in \mathbb{N}_0} |2^{sj} u_j^*(\vec{b}^j; t, x)|$ is integrable. Hence, Lebesgue's dominated convergence theorem implies that $\sum_{j=0}^{k} u_j(\cdot, z)$ is a Cauchy sequence in $L_{q,\vec{p}'}(G_{n-1})$ and thus is convergent. Since every partial sum satisfies the estimate by (3.35) so does the limit. Therefore, the results follows by Proposition 3.4.1 and (3.36). $\qquad\square$

As a special case we collect the case of $z = 0$ in the following corollary.

Corollary 3.4.3. *The series defining $T_n u$ converges in $L_{q,\vec{p}'}(G_{n-1})$. The trace operator T_n is continuous from $F^{s,(b,\vec{a})}_{(q,\vec{p}),r}(G_n)$ to $L_{q,\vec{p}'}(G_{n-1})$ and satisfies the estimate*

$$(3.38) \qquad \|T_n u\|_{L_{q,\vec{p}'}(G_{n-1})} = \Big\| \sum_{j=0}^{\infty} u_j(\cdot, 0) \Big\|_{L_{q,\vec{p}'}(G_{n-1})} \lesssim \|u\|_{F^{s,(b,\vec{a})}_{(q,\vec{p}),r}(G_n)}$$

for all $u \in F^{s,(b,\vec{a})}_{(q,\vec{p}),r}(G_n)$.

The natural question is now whether the operator T_n coincides with the classical trace. This is indeed the case as shown by the following lemma.

Lemma 3.4.4. *For $s > \frac{a_n}{p_n}$ the embedding $F^{s,(b,\vec{a})}_{(q,\vec{p}),r}(G_n) \hookrightarrow C(\mathbb{R}, L_{q,\vec{p}'}(G_{n-1}))$ is continuous and $T_n u = u(\cdot, 0)$ holds.*

CHAPTER 4

The Equations of Magnetohydrodynamics

In this chapter, we consider the equations of magnetohydrodynamics and prove existence of a time-periodic solution. Therefore, we recall the system (MHDE) derived in the introduction given by

(MHD)
$$
\begin{cases}
\partial_t u - \Delta u + \nabla \mathrm{p} + \dfrac{1}{2}\nabla|H|^2 + (u \cdot \nabla)u = F + (H \cdot \nabla)H & \text{in } \Omega_{\mathbb{T}}, \\[2mm]
\partial_t H - \Delta H = \nabla \times [u \times H] & \text{in } \Omega_{\mathbb{T}}, \\[2mm]
\operatorname{div} u = \operatorname{div} H = 0 & \text{in } \Omega_{\mathbb{T}}, \\[2mm]
u = 0, \quad H \cdot n = B_1, \quad \operatorname{curl} H \times n = 0 & \text{on } \partial\Omega_{\mathbb{T}}.
\end{cases}
$$

The function F is a given time-periodic function on $\Omega_{\mathbb{T}} = \mathbb{T} \times \Omega$, where $\Omega \subset \mathbb{R}^3$ is a bounded simply connected domain with outer normal vector n and boundary of class $C^{2,1}$. The constants μ and ν were set to 1 to simplify the notation and it is readily checked that at no point the actual values are of importance, therefore the restriction is without loss of generality. As Yoshida and Giga [87] pointed out and we briefly remarked in the introduction the inhomogeneity B_1 is a given time-independent function which corresponds to the initial magnetic field of the medium containing the liquid. Since it corresponds to a magnetic field it must satisfy $\int_{\partial\Omega} B_1(x)\, \mathrm{d}\sigma = 0$. Otherwise it could not satisfy the last equation in the Maxwell equations, see (ME), by the validity of the divergence theorem.

Since we want to be able to deal with arbitrary large background magnetic fields B_1, see Remark 4.4.5, we construct an extension of B_1. Therefore, we consider the inhomogeneous Neumann problem of the Laplace operator given by

(4.1)
$$
\begin{cases}
\Delta v = 0 & \text{in } \Omega, \\[2mm]
\nabla v \cdot n = B_1 & \text{on } \partial\Omega.
\end{cases}
$$

This problem can be considered independent of $t \in \mathbb{T}$, since B_1 does not depend on time. It is known that for $1 < r < \infty$ and every $B_1 \in \mathrm{W}_r^{1-\frac{1}{r}}(\partial\Omega)$ that fulfils $\int_{\partial\Omega} B_1(x)\, \mathrm{d}x = 0$, there exists a solution $v \in \mathrm{W}_r^2(\Omega)$ to (4.1), see for example Amann [5, Remark 7.3 and Remark 9.5] or Seyfert [76, Lemma 1.4.3]. By assuming $B_1 \in \mathrm{W}_r^{2-\frac{1}{r}}(\partial\Omega)$ we derive from standard theory, see for example Grisvard [42, Theorem 2.5.1.1] , that the solution v is an element of $W_r^3(\Omega)$ that satisfies the estimate

$$(4.2) \qquad \|v\|_{W_\infty^2(\Omega)} \lesssim \|v\|_{W_r^3(\Omega)} \lesssim \|B_1\|_{W_r^{2-\frac{1}{r}}(\partial\Omega)}$$

for $r > 3$, where the first estimate is an application of Sobolev embeddings and requires $r > 3$. We set $H_0 = \nabla v$ and conclude $\nabla \times H_0 = 0$, $\nabla \cdot H_0 = 0$. Hence H_0 is a solenoidal vector field that satisfies the boundary conditions and is an element of $W_\infty^1(\Omega) \cap W_r^2(\Omega)$. Furthermore, since weak derivatives always commute we have $\Delta H_0 = \Delta \nabla v = \nabla \Delta v = 0$. From here on we fix H_0 as the element obtained in the described way and transform the equations (MHD) via $H = \tilde{H} + H_0$ to derive

$$\begin{cases} \partial_t u - \Delta u - (H_0 \cdot \nabla)\tilde{H} - (\tilde{H} \cdot \nabla)H_0 + \nabla\mathfrak{p} = F + (H_0 \cdot \nabla)H_0 + (\tilde{H} \cdot \nabla)\tilde{H} - (u \cdot \nabla)u & \text{in } \Omega_{\mathbb{T}}, \\ \partial_t \tilde{H} - \Delta \tilde{H} - \nabla \times [u \times H_0] = \nabla \times [u \times \tilde{H}] & \text{in } \Omega_{\mathbb{T}}, \\ \operatorname{div} u = \operatorname{div} \tilde{H} = 0 & \text{in } \Omega_{\mathbb{T}}, \\ u = 0, \quad \tilde{H} \cdot n = 0, \quad \operatorname{curl} \tilde{H} \times n = 0 & \text{on } \partial\Omega_{\mathbb{T}}, \end{cases}$$

where we defined the new pressure term $\mathfrak{p} := p + \frac{1}{2}|\tilde{H} + H_0|^2$. As a first step we are going to consider the linearization of the transformed equations. For simplicity of notation, since there is no confusion possible, we will denote the unknown functions still by H and hence the equations are given by

$$(4.3) \qquad \begin{cases} \partial_t u - \Delta u - (H_0 \cdot \nabla)H - (H \cdot \nabla)H_0 + \nabla\mathfrak{p} = F & \text{in } \Omega_{\mathbb{T}}, \\ \partial_t H - \Delta H - \nabla \times [u \times H_0] = G & \text{in } \Omega_{\mathbb{T}}, \\ \operatorname{div} u = \operatorname{div} H = 0 & \text{in } \Omega_{\mathbb{T}}, \\ u = 0, \quad H \cdot n = 0, \quad \operatorname{curl} H \times n = 0 & \text{on } \partial\Omega_{\mathbb{T}}. \end{cases}$$

Readers familiar with fluid dynamics might notice, that this system seems to be overdetermined. In Section 4.1 we will show that this is not the case, because the Helmholtz projection commutes with ΔH under these boundary conditions and that $G \in L_{q,p,\sigma}(\Omega)$ can be assumed. Therefore, the system is well-posed. Furthermore, we will show that the additional terms from the extension of the boundary data can be viewed as a perturbation and that the perturbed system still has a resolvent set containing the whole imaginary axis. The last point is of importance for the considerations in Section 4.3.

Before we are able to make use of the perturbation results of Section 4.1 we need a good understanding of the unperturbed equations. In the case of the Stokes equations, *i.e.*, the equations concerning the velocity u, the linear theory can be found in Maekawa and Sauer [66]. Therefore, the unperturbed system for H will be considered in Section 4.2. This will be the study of the time-periodic heat equation with the perfect conductivity boundary condition, that is, we consider the equations

$$(4.4) \qquad \begin{cases} \partial_t H - \mu\Delta H = G & \text{in } \Omega_{\mathbb{T}}, \\ \operatorname{div} H = 0 & \text{in } \Omega_{\mathbb{T}}, \\ H \cdot n = 0 \quad \operatorname{curl} H \times n = 0 & \text{on } \partial\Omega_{\mathbb{T}}, \end{cases}$$

for a given function $G \in \mathcal{P}_\perp L_{q,p,\sigma}(\Omega_{\mathbb{T}})$. Here we only need to consider the purely periodic part, since the stationary part was already considered, see Section 1.8. Furthermore, in Section 4.2 we will see the advantage of splitting the problem via the projections $\mathcal{P}$ and $\mathcal{P}_\perp$ and we will make very frequent use of the theory and results developed in Chapter 3.

Section 4.2 together with the work of Maekawa and Sauer [66] allows us to apply the results of Section 4.1 to (4.3) to conclude existence of solutions and maximal regularity in the $\mathrm{L}_{q,p}(\Omega_\mathrm{T})$-framework. It is worth noting that one cannot expect to obtain estimates with a constant completely independent of H_0, but we will show that if the $\mathrm{W}^1_\infty(\Omega)$-norm of H_0 can be controlled by a constant, then these estimates will be uniform for all H_0 satisfying the norm estimate.

In the last section we will combine all of the previous results and prove the main result of this chapter, the existence of a time-periodic solution to (MHD), by an application of Banach's fixed-point theorem.

Recall that we use the same notation for scalar-valued and vector-valued functions, unless it might cause confusion.

4.1 Properties of the Boundary Conditions and Operator Theory

As a first step we use the Helmholtz projection to transform the equations in (4.3) into equations without a pressure term. It is known that for u the projection does not commute with Δu. Since H satisfies different boundary conditions we will show that in this case the operators commute.

Lemma 4.1.1. *Let $1 < p < \infty$ and $\Omega \subset \mathbb{R}^3$ a bounded domain of class C^2. For any function $H \in \mathrm{W}^2_p(\Omega)$ that satisfies $\operatorname{div} H = 0$ and $\operatorname{curl} H \times n = 0$ we have $\Delta H \in \mathrm{L}_{p,\sigma}(\Omega)$. Furthermore, for $u \in \mathrm{W}^2_p(\Omega)$ such that $u = 0$ on $\partial\Omega$ and $H \in \mathrm{W}^2_p(\Omega)$ it holds $\operatorname{curl}(u \times H) \in \mathrm{L}_{p,\sigma}(\Omega)$.*

Proof: By (1.11) we need to show

$$\int_\Omega \Delta H \cdot \nabla\varphi \, \mathrm{d}x = 0$$

for all $\varphi \in \mathrm{W}^1_{p'}(\Omega)$. Since $\mathrm{C}^\infty_0(\overline{\Omega})$ is dense in $\mathrm{W}^1_{p'}(\Omega)$ it suffices to show equality for these smooth functions. By $\Delta H = \nabla(\operatorname{div} H) - \operatorname{curl}\operatorname{curl} H$ we obtain $\Delta H = -\operatorname{curl}\operatorname{curl} H$. The divergence theorem with respect to curl and a vector field ψ has the form

$$\int_{\partial\Omega} (H \times \psi) \cdot n \, \mathrm{d}\sigma = \int_\Omega (\operatorname{curl} H) \cdot \psi - H \cdot (\operatorname{curl}\psi) \, \mathrm{d}x.$$

Thus

$$\int_\Omega \Delta H \cdot \nabla\varphi \, \mathrm{d}x = -\int_\Omega \operatorname{curl}\operatorname{curl} H \cdot \nabla\varphi \, \mathrm{d}x = -\int_\Omega \operatorname{curl} H \cdot \operatorname{curl} \nabla\varphi \, \mathrm{d}x - \int_{\partial\Omega} (\operatorname{curl} H \times \nabla\varphi) \cdot n \, \mathrm{d}\sigma$$

$$\text{(4.5)} \qquad = \int_{\partial\Omega} (\operatorname{curl} H \times n) \cdot \nabla\varphi \, \mathrm{d}\sigma = 0.$$

Before we start with the second case, we note that by $\nabla \times [u \times H] = (H \cdot \nabla)u - (u \cdot \nabla)H$ and Lemma 4.4.1 below, which is applicable since we can pick q arbitrarily large, $\nabla \times [u \times H]$ is an element of $\mathrm{L}_p(\Omega)$ and therefore all of the following integrals exists. It holds

$$\int_\Omega \operatorname{curl}(u \times H) \cdot \nabla\varphi \, \mathrm{d}x = \int_\Omega (u \times H) \cdot \operatorname{curl}(\nabla\varphi) \, \mathrm{d}x + \int_{\partial\Omega} (u \times H \times \nabla\varphi) \cdot n \, \mathrm{d}\sigma = 0,$$

since u vanishes on the boundary and $\operatorname{curl} \nabla\varphi = 0$. $\qquad\square$

With the result of the previous Lemma we apply the Helmholtz projection to the equations (4.3) to obtain the reformulation

(4.6)
$$\begin{cases} \partial_t u - \mathcal{P}_H \Delta u - \mathcal{P}_H (H_0 \cdot \nabla)H - \mathcal{P}_H (H \cdot \nabla)H_0 = \mathcal{P}_H F & \text{in } \Omega_T, \\ \partial_t H - \Delta H + (u \cdot \nabla)H_0 - (H_0 \cdot \nabla)u = G & \text{in } \Omega_T, \\ \operatorname{div} u = \operatorname{div} H = 0 & \text{in } \Omega_T, \\ u = 0, \quad H \cdot n = 0, \quad \operatorname{curl} H \times n = 0 & \text{on } \partial\Omega_T. \end{cases}$$

Here we used that $G \in L_{q,p,\sigma}(\Omega_T)$ by Lemma 4.1.1 and the identity $\nabla \times [u \times H_0] = (H_0 \cdot \nabla)u - (u \cdot \nabla)H_0$ for solenoidal vector fields u and H. An important tool when working with the Laplace operator is the fact that it commutes with rotations, *i.e.*, $\Delta[u(Qx)] = (\Delta u)(Qx)$ for any rotation matrix Q. In the following we are going to show that the boundary conditions of H are invariant with respect to rotations, which will allow us to use rotations in Section 4.2.2.

Lemma 4.1.2. *For every $a, b \in \mathbb{R}^3$ and rotation matrix $Q \in \mathbb{R}^{3\times3}$ we have*

$$Q(a \times b) = Qa \times Qb.$$

Proof: Since Q is a rotation matrix we have $Q^\mathsf{T}Q = \operatorname{Id}$. Note that with Einstein's summation convention we conclude

$$Qa \times Qb = \begin{pmatrix} q_{1j}a_j \\ q_{2j}a_j \\ q_{3j}a_j \end{pmatrix} \times \begin{pmatrix} q_{1j}b_j \\ q_{2j}b_j \\ q_{3j}b_j \end{pmatrix} = \begin{pmatrix} q_{2j}a_j q_{3k}b_k - q_{3j}a_j q_{2k}b_k \\ q_{3j}a_j q_{1k}b_k - q_{1j}a_j q_{3k}b_k \\ q_{1j}a_j q_{2k}b_k - q_{2j}a_j q_{1k}b_k \end{pmatrix}$$

and hence $Q^\mathsf{T}(Qa \times Qb)$ is equal to

$$\begin{pmatrix} q_{11}(q_{2j}a_j q_{3k}b_k - q_{3j}a_j q_{2k}b_k) + q_{21}(q_{3j}a_j q_{1k}b_k - q_{1j}a_j q_{3k}b_k) + q_{31}(q_{1j}a_j q_{2k}b_k - q_{2j}a_j q_{1k}b_k) \\ q_{12}(q_{2j}a_j q_{3k}b_k - q_{3j}a_j q_{2k}b_k) + q_{22}(q_{3j}a_j q_{1k}b_k - q_{1j}a_j q_{3k}b_k) + q_{32}(q_{1j}a_j q_{2k}b_k - q_{2j}a_j q_{1k}b_k) \\ q_{13}(q_{2j}a_j q_{3k}b_k - q_{3j}a_j q_{2k}b_k) + q_{23}(q_{3j}a_j q_{1k}b_k - q_{1j}a_j q_{3k}b_k) + q_{33}(q_{1j}a_j q_{2k}b_k - q_{2j}a_j q_{1k}b_k) \end{pmatrix}.$$

Considering each component separately, we see that

$$[Q^\mathsf{T}(Qa \times Qb)]_i = a_j b_k (q_{1i}q_{2j}q_{3k} - q_{1i}q_{3j}q_{2k} + q_{2i}q_{3j}q_{1k} - q_{2i}q_{1j}q_{3k} + q_{3i}q_{1j}q_{2k} - q_{3i}q_{2j}q_{1k}).$$

Hence it follows that for $i = k$, $i = j$ or $j = k$ the term in brackets vanishes. So we are left to deal with the case of pairwise disjoint i, j, k. Furthermore, we see that the sum on the right-hand side is $\pm \det Q$, depending on the choice of i, j, k, as we sum over all permutations of i, j, k and the sign of each term is consistent with the order of permutation with respect to the rest. Hence, it suffices to check the first summand to determine the sign, thus we obtain

$$Q^\mathsf{T}(Qa \times Qb) = a \times b,$$

which proves the result. $\qquad\square$

For the next step we consider domains Ω and $\tilde{\Omega}$ such that $Q(\Omega) = \tilde{\Omega}$ for a rotation matrix Q. For a vector field $u : \Omega \to \mathbb{R}^3$ we define

$$(4.7) \qquad \Phi u : \tilde{\Omega} \to \mathbb{R}^3, \quad \text{where} \quad \Phi u(x) := Q^\mathsf{T} u(Qx).$$

A simple consequence of the previous Lemma is the following.

Lemma 4.1.3. *For every rotation matrix $Q \in \mathbb{R}^{3\times3}$ we have the identity*

$$\mathrm{curl}[\Phi(u)] = \Phi[\mathrm{curl}\, u].$$

Proof: It holds

$$\mathrm{curl}[\Phi(u)] = \begin{pmatrix} q_{j2}q_{k3}\partial_j u_k - q_{k2}q_{j3}\partial_j u_k \\ q_{k1}q_{j3}\partial_j u_k - q_{j1}q_{k3}\partial_j u_k \\ q_{j1}q_{k2}\partial_j u_k - q_{k1}q_{j2}\partial_j u_k \end{pmatrix} \circ Q$$

and therefore

$$Q\,\mathrm{curl}[\Phi(u)] = (\mathrm{curl}\, u) \circ Q$$

by repeating the calculations made in Lemma 4.1.2. $\qquad\square$

With this preparations we can show that Φ preserves the boundary condition of H of equations (4.6).

Lemma 4.1.4. *The boundary conditions $H \cdot n = 0$ and $\mathrm{curl}\, H \times n = 0$ are invariant under the transformation Φ.*

Proof: Note that, since $\mathrm{cof}(Q) = Q$, the transformed outer normal vector is given by $n \circ Q = Q\tilde{n}$, where n is the outer normal of Ω and $\tilde{n}$ of $\tilde{\Omega}$. Hence, we conclude

$$\Phi(H) \cdot \tilde{n} = Q^\mathsf{T} H(Qx) \cdot \tilde{n} = H(Qx) \cdot Q\tilde{n} = [H(x) \cdot n] \circ Q,$$
$$\mathrm{curl}\, \Phi(H) \times \tilde{n} = [Q^\mathsf{T} \mathrm{curl}\, H \circ Q] \times \tilde{n} = [Q^\mathsf{T} \mathrm{curl}\, H \circ Q] \times Q^\mathsf{T} n \circ Q = Q^\mathsf{T}[(\mathrm{curl}\, H \times n) \circ Q]$$
$$= \Phi(\mathrm{curl}\, H \times n),$$

which completes the proof. $\qquad\square$

To be able to apply perturbation theory, we introduce the following operators

$$A_p := -\mathcal{P}_H \Delta, \quad D(A_p) := \mathrm{W}_p^2(\Omega) \cap \mathrm{W}_{p,0}^1(\Omega) \cap \mathrm{L}_{p,\sigma}(\Omega),$$
$$B_p := -\Delta, \quad D(B_p) := \{ H \in \mathrm{W}_p^2(\Omega) \mid H \cdot n = 0 \text{ and } \mathrm{rot}\, H \times n = 0 \text{ on } \partial\Omega \} \cap \mathrm{L}_{p,\sigma}(\Omega),$$

where $\mathrm{W}_{p,0}^1(\Omega) := \{ u \in \mathrm{W}_p^1(\Omega) \mid u = 0 \text{ on } \partial\Omega \}$. It is well-known that the Stokes operator A_p generates an analytic bounded semi-group on $\mathrm{L}_{p,\sigma}(\Omega)$ for $1 < p < \infty$. By the results of Section 1.8 the same holds for B_p on $\mathrm{L}_{p,\sigma}(\Omega)$ for $1 < p < \infty$, hence the domains equipped with the graph norm are Banach spaces. Furthermore, it is well-known for the Stokes operator that $A_p^* = A_{p'}$. The same holds for B_p, i.e, $B_p^* = B_{p'}$, see Al Baba, Amrouche and Escobedo [4, Section 3.2.1]. We start with a continuous embedding result regarding the domains.

Lemma 4.1.5. *For $1 < p < \infty$ we get the continuous embeddings*

$$\left[L_p(\Omega), D(A_p)\right]_{\frac{1}{2}} \hookrightarrow \left[L_p(\Omega), W_p^2(\Omega)\right]_{\frac{1}{2}} = W_p^1(\Omega),$$

$$\left[L_p(\Omega), D(B_p)\right]_{\frac{1}{2}} \hookrightarrow \left[L_p(\Omega), W_p^2(\Omega)\right]_{\frac{1}{2}} = W_p^1(\Omega).$$

Proof: The identity of the complex interpolation space with W_p^1 follows from standard interpolation theory. The space $W_p^2(\Omega) \cap W_{p,0}^1(\Omega) \cap L_{p,\sigma}(\Omega)$ equipped with the norm $\|\cdot\|_{W_p^2(\Omega)}$ is a Banach space. Therefore, we consider the natural injection $\iota : W_p^2(\Omega) \cap W_{p,0}^1(\Omega) \cap L_{p,\sigma}(\Omega) \hookrightarrow D(A_p)$ which satisfies the estimate

$$\|\iota(u)\|_{D(A_p)} = \|A_p u\|_{L_p(\Omega)} + \|u\|_{L_p(\Omega)} \lesssim \|u\|_{W_p^2(\Omega)}.$$

Thus ι is a continuous bijective operator and hence ι^{-1} is continuous by the open mapping theorem. This implies that the embedding $D(A_p) \hookrightarrow W_p^2(\Omega)$ is continuous, which implies the first result. The second follows by the same ideas since $\{u \in W_p^2(\Omega) \mid u \cdot n = 0 \text{ and } \operatorname{rot} u \times n = 0 \text{ on } \partial\Omega\}$ is a closed subset of $W_p^2(\Omega)$ by equations (1.9) and (1.37). $\qquad\square$

By using the introduced operators, we rewrite (4.6) into an operator setting. It holds

$$\partial_t \begin{pmatrix} u \\ H \end{pmatrix} + \begin{pmatrix} A_p & 0 \\ 0 & B_p \end{pmatrix} \begin{pmatrix} u \\ H \end{pmatrix} + \begin{pmatrix} -\mathcal{P}_H(H_0 \cdot \nabla H) - \mathcal{P}_H(\nabla H_0 \cdot H) \\ -\nabla u \cdot H_0 + \nabla H_0 \cdot u \end{pmatrix} = \begin{pmatrix} \mathcal{P}_H F \\ G \end{pmatrix}.$$

We want to make use of perturbation results regarding closed operators and hence we define

$$(4.8) \qquad \begin{aligned} S_p^1 &: D(B_p) \to L_p(\Omega), \quad S_p^1(H) := -\nabla H \cdot H_0 - \nabla H_0 \cdot H, \\ S_p^2 &: D(A_p) \to L_{p,\sigma}(\Omega), \quad S_p^2(u) := -\nabla u \cdot H_0 + \nabla H_0 \cdot u. \end{aligned}$$

The operator corresponding to our problem will be denoted by T_p and is defined by

$$(4.9) \qquad T_p := (A_p, B_p) + (\mathcal{P}_H S_p^1, S_p^2) : D(A_p) \times D(B_p) \to L_{p,\sigma}(\Omega) \times L_{p,\sigma}(\Omega).$$

For the next lemma recall that $H_0 \in W_\infty^1(\Omega)$ by assumption.

Lemma 4.1.6. *The operator T_p from (4.9) is a closed operator with compact resolvent and there exists $\lambda \in (-\infty, 0)$ in its resolvent set. Furthermore, the spectrum only consists of eigenvalues for all $p \in (1, \infty)$.*

Proof: By applying Lemma 4.1.5 together with the interpolation inequality for complex interpolation we obtain

$$(4.10) \qquad \begin{aligned} \left\|(\mathcal{P}_H S_p^1, S_p^2)\right\|_{L_p} &\lesssim \|H\|_{W_p^1(\Omega)} + \|u\|_{W_p^1(\Omega)} + \lesssim \|H\|_{L_p}^{\frac{1}{2}} \|H\|_{D(B_p)}^{\frac{1}{2}} + \|u\|_{L_p}^{\frac{1}{2}} \|u\|_{D(A_p)}^{\frac{1}{2}} \\ &\lesssim C(\varepsilon) \|(u, H)\|_{L_p} + \varepsilon \|(A_p u, B_p H)\|_{L_p} \end{aligned}$$

Hence the operator $(\mathcal{P}_H S_p^1, S_p^2)$ is relatively bounded with respect to (A_p, B_p) and relative bound 0. Furthermore, with resolvent $R(\lambda, A)$ defined by $R(\lambda, A) = (A - \lambda)^{-1}$, we derive an $M > 0$ such that

$$\|\lambda R(\lambda, B_p)\|_{\mathcal{L}(L_p)} + \|\lambda R(\lambda, A_p)\|_{\mathcal{L}(L_p)} < M, \quad \text{for all } \lambda \in (-\infty, 0),$$

since A_p and B_p are sectorial operators. Hence it holds

$$C(\varepsilon)\|R(\lambda, A_p)\|_{\mathcal{L}(\mathrm{L}_p)} + \varepsilon\|A_pR(\lambda, A_p)\|_{\mathcal{L}(\mathrm{L}_p)} \leq \frac{C(\varepsilon)M}{|\lambda|} + \varepsilon(M+1) < \frac{1}{2}$$

$$C(\varepsilon)\|R(\lambda, B_p)\|_{\mathcal{L}(\mathrm{L}_p)} + \varepsilon\|B_pR(\lambda, B_p)\|_{\mathcal{L}(\mathrm{L}_p)} \leq \frac{C(\varepsilon)M}{|\lambda|} + \varepsilon(M+1) < \frac{1}{2}$$

for fixed ε and sufficiently large $|\lambda|$. This implies by Kato [57, Theorem IV.3.17] that T_p is a closed operator with compact resolvent, since A_p and B_p have compact resolvents, and all λ satisfying the above estimates are in the resolvent. Hence Engel and Nagel [31, Corollary IV.1.19] implies that T_p has a spectrum only consisting of eigenvalues for all $p \in (1, \infty)$. $\qquad\square$

With this preparation we can show that the imaginary line is contained in the resolvent set of T_p for all values of $p \in (1, \infty)$. As we have seen in Section 1.8 the Stokes operator with Navier type boundary conditions is not invertible in a general domain and we cannot expect that the perturbed operator possesses better properties. Therefore, the restrictions made to consider simply connected domains are vital for the next result.

Theorem 4.1.7. *Let $\Omega \subset \mathbb{R}^3$ be a simply connected bounded domain of class $\mathrm{C}^{2,1}$ and $1 < p < \infty$. Then there exists some $\delta > 0$ such that the set $\{z \in \mathbb{C} \mid \mathrm{Re}\, z < \delta\}$ is contained in the resolvent set of T_p for every $p \in (1, \infty)$ if $\nabla H_0 = (\nabla H_0)^{\mathsf{T}}$.*

Proof: We first consider $\mathrm{L}_{2,\sigma}$, meaning we consider T_2, and recall that the numerical range is defined by

$$\Theta(T) := \{\langle Tx, \overline{x}\rangle_{\mathrm{L}_{2,\sigma} \times \mathrm{L}_{2,\sigma}} \mid x \in \mathrm{D}(T), \|x\|_{\mathrm{L}_{2,\sigma} \times \mathrm{L}_{2,\sigma}} = 1\},$$

where $\langle \cdot, \cdot \rangle$ denotes the L_2 scalar product. For $u \in D(A_2)$ and $H \in D(B_2)$ it holds

$$\left\langle \begin{pmatrix} A_2u - \mathcal{P}_H(\nabla H \cdot H_0) - \mathcal{P}_H(\nabla H_0 \cdot H) \\ B_2H - \nabla u \cdot H_0 + \nabla H_0 \cdot u \end{pmatrix}, \begin{pmatrix} \overline{u} \\ \overline{H} \end{pmatrix} \right\rangle$$

$$= \|\nabla u\|_{\mathrm{L}_2}^2 + \|\nabla \times H\|_{\mathrm{L}_2}^2 - \int_\Omega (\nabla H \cdot H_0)\overline{u} + (\nabla H_0 \cdot H)\overline{u} + (\nabla u \cdot H_0)\overline{H} - (\nabla H_0 \cdot u)\overline{H} \, dx$$

$$= \|\nabla u\|_{\mathrm{L}_2}^2 + \|\nabla \times H\|_{\mathrm{L}_2}^2 + \int_\Omega (\nabla \overline{u} \cdot H_0)H - (\nabla H_0 \cdot H)\overline{u} - (\nabla u \cdot H_0)\overline{H} + (\nabla H_0 \cdot u)\overline{H} \, dx$$

$$= \|\nabla u\|_{\mathrm{L}_2}^2 + \|\nabla \times H\|_{\mathrm{L}_2}^2 + \int_\Omega (\nabla H_0 \cdot \mathrm{Re}\, u)\,\mathrm{Re}\, H - (\nabla H_0 \cdot \mathrm{Re}\, H)\,\mathrm{Re}\, u \, dx$$

$$+ \int_\Omega (\nabla H_0 \cdot \mathrm{Im}\, u)\,\mathrm{Im}\, H - (\nabla H_0 \cdot \mathrm{Im}\, H)\,\mathrm{Im}\, u \, dx$$

$$+ i\int_\Omega (\nabla H_0 \cdot \mathrm{Im}\, u)\,\mathrm{Re}\, H - (\nabla H_0 \cdot \mathrm{Re}\, u)\,\mathrm{Im}\, H + (\nabla H_0 \cdot \mathrm{Re}\, H)\,\mathrm{Im}\, u - (\nabla H_0 \cdot \mathrm{Im}\, H)\,\mathrm{Re}\, u \, dx$$

$$+ 2i\int_\Omega (\nabla \mathrm{Re}\, u \cdot H_0)\,\mathrm{Im}\, H - (\nabla \mathrm{Im}\, u \cdot H_0)\,\mathrm{Re}\, H \, dx$$

Since $\nabla H_0 = (\nabla H_0)^{\mathsf{T}}$, we derive $(\nabla H_0 \cdot \mathrm{Re}\, u)\,\mathrm{Re}\, H - (\nabla H_0 \cdot \mathrm{Re}\, H)\,\mathrm{Re}\, u = 0$, and the same holds for the other real-valued expression. Because u and $H \cdot n$ are equal to 0 on $\partial\Omega$ inequality (1.35) implies that the sum of the L_2-norms of ∇u and $\mathrm{curl}\, H$ are strictly away from 0. Therefore, there exists an $\delta > 0$ such that the numerical range is contained in the set $\{z \in \mathbb{C} \mid \mathrm{Re}\, z > \delta\}$. We derive from Lemma 4.1.6 that the operator T_2 is closed and $(T_2 - \lambda)$ is invertible for some $\lambda < -1$. Hence Kato [57, Theorem V.3.2] implies that the set $\{z \in \mathbb{C} \mid \operatorname{Re} z < \delta\}$ is in the resolvent set of T_2. Next we want to show that this holds for all $p \in (1, \infty)$. As we only have to deal with eigenfunctions it is clear that for $p > 2$ the result is true, since it holds $\mathrm{W}_p^2(\Omega) \hookrightarrow \mathrm{W}_2^2(\Omega)$. For $1 < p < 2$ we consider the adjoint operator. If we take $u \in D(A_p)$, $H \in D(B_p)$ and $v \in D(A_{p'})$, $M \in D(B_{p'})$ we obtain from the calculations above, $\nabla H_0 = (\nabla H_0)^\mathsf{T}$, and using the fact that the dual operator of $\mathcal{P}_H$ on L_p is $\mathcal{P}_H$ on $\mathrm{L}_{p'}$ that

$$\left\langle \begin{pmatrix} -\mathcal{P}_H(\nabla H \cdot H_0) - \mathcal{P}_H(\nabla H_0 \cdot H) \\ -\nabla u \cdot H_0 + \nabla H_0 \cdot u \end{pmatrix}, \begin{pmatrix} v \\ M \end{pmatrix} \right\rangle = \left\langle \begin{pmatrix} u \\ H \end{pmatrix}, \begin{pmatrix} \mathcal{P}_H(\nabla M \cdot H_0) + \mathcal{P}_H(\nabla H_0 \cdot M) \\ \nabla v \cdot H_0 - \nabla H_0 \cdot v \end{pmatrix} \right\rangle.$$

This shows that $D(A_{p'}) \times D(B_{p'})$ is contained in the domain of T_p^*. Since $(A_p)^* = A_{p'}$ and $(B_p)^* = B_{p'}$ we conclude by the same calculations as in (4.10) that $([\mathcal{P}_H S_p^1]^*, [S_p^2]^*)$ is relatively bounded with respect to $(A_p^*, B_p^*) = (A_{p'}', B_{p'}')$ with relative bound 0. Because on $D(A_{p'}) \times D(B_{p'})$ it has the same form as T_p but with $-H_0$ instead of H_0. Hence Hess and Kato [44, Corollary 1] imply that the operator calculated above is indeed the adjoint of T_p. By our calculations from before we see that the choice of H_0 has no influence on the existence of $\delta > 0$, as long as it satisfies $\nabla H_0 = (\nabla H_0)^\mathsf{T}$. Therefore, the set $\{z \in \mathbb{C} \mid \operatorname{Re} z < \delta\}$ is contained in the resolvent set of T_p^* for $p > 2$ and all H_0 with the stated property. Since the resolvent set of T_p is just the resolvent set of T_p^* mirrored at the real axis, see Kato [57, Theorem III.6.22], we conclude the results for all $1 < p < \infty$. $\qquad \square$

4.2 Time-Periodic Solutions to the Heat Equation with Perfect Conductivity Boundary Conditions

In this section, we will deal with the question of time-periodic maximal regularity of the following differential equations

$$(4.11) \qquad \begin{cases} \partial_t H - \mu \Delta H = f & \text{in } \Omega_\mathbb{T}, \\ \operatorname{div} H = 0 & \text{in } \Omega_\mathbb{T}, \\ H \cdot n = 0 \quad \operatorname{curl} H \times n = 0 & \text{on } \partial\Omega_\mathbb{T}, \end{cases}$$

where Ω is a bounded domain of class $C^{2,1}$, not necessarily simply connected, with outer normal vector n, $f \in \mathcal{P}_\perp \mathrm{L}_{q,p,\sigma}(\Omega_\mathbb{T})$ and $\mu \in \mathbb{R} \setminus \{0\}$ are given, whereas $H : \Omega \to \mathbb{R}^3$ is the unknown function. The function f is given as a purely periodic function, since the stationary part was investigated in Section 1.8. We start our consideration in the whole and half space and recall that $G_n = \mathbb{T} \times \mathbb{R}^n$.

4.2.1 The Whole and the Half Space

In the case of the whole or half spaces we study the slightly modified problem and consider

$$(4.12) \qquad \begin{cases} \partial_t H - \mu \Delta H = f & \text{in } \Omega_\mathbb{T}, \\ H \cdot n = 0 \quad \operatorname{curl} H \times n = g & \text{on } \partial\Omega_\mathbb{T}, \end{cases}$$

where $\Omega = \mathbb{R}^3$ or $\mathbb{R}_+^3$, and f and g are given functions. In the case of $\mathbb{R}^3$ we omit the boundary data and briefly prove existence of a solution and suitable estimates. The proof relies on multiplier theory and is in principle known, but since we are dealing with different powers of integrability with respect to time and space we give a short proof.

Lemma 4.2.1. *Let $1 < q, p < \infty$, $\mu \in \mathbb{R} \setminus \{0\}$, and $f \in \mathcal{P}_\perp L_{q,p}(G_3)$. Then there exists a unique solution H to (4.12) in $\mathcal{P}_\perp W^{1,2}_{q,p}(G_3)$ and a constant $c > 0$ such that*

$$(4.13) \qquad \|H\|_{W^{1,2}_{q,p}(G_3)} \leq c\|f\|_{L_{q,p}(G_3)}$$

holds. If $f \in \mathcal{P}_\perp L_{q,p,\sigma}(G_3)$ then the solution H satisfies $\operatorname{div} H = 0$.

Proof: The result follows from $m(k, \xi) = \frac{1-\delta_0(k)}{ik+\mu|\xi|^2}$ being an $L_{q,p}(G_3)$-multiplier, the same holds if it is multiplied by ik or ξ^α for $|\alpha| \leq 2$. Details can be found in Kyed and Sauer [61, Lemma 5.2]. Their arguments combined with Theorem 2.2.5 and Theorem 1.3.1 yield a solution that satisfies (4.13). Since all $f \in L_{q,p,\sigma}(G_3)$ satisfy $\mathscr{F}^{-1}[\xi \cdot \mathscr{F}f] = 0$ the same holds for H and hence $\operatorname{div} H = 0$.

Suppose that there are two solutions H and B in $\mathcal{P}_\perp W^{1,2}_{q,p}(G_3)$. The difference $H - B$ satisfies $(ik + \mu|\xi|^2)\mathscr{F}(H - B) = 0$ and therefore $\operatorname{supp} \mathscr{F}(H - B) \subset \{(0,0)\}$. Because $0 = \mathcal{P}H = \mathcal{P}B$ we obtain $0 = \mathscr{F}[\mathcal{P}(H - B)] = \delta_0 \cdot \mathscr{F}(H - B)$ and thus $H = B$ since $(0,0) \notin \operatorname{supp} \mathscr{F}[H - B]$. $\square$

Before we consider the half space we briefly show identities regarding the boundary conditions. The outer normal vector to $\mathbb{R}^3_+$ is given by $-e_3$ and it holds

$$(4.14) \qquad H\big|_{\partial \mathbb{R}^3_+} \cdot n = 0 \iff H_3\big|_{\partial \mathbb{R}^3_+} = 0,$$

$$(4.15) \qquad \operatorname{curl} H\big|_{\partial \mathbb{R}^3_+} \times n = \begin{pmatrix} \partial_2 H_3 - \partial_3 H_2 \\ \partial_3 H_1 - \partial_1 H_3 \\ \partial_1 H_2 - \partial_2 H_1 \end{pmatrix} \times \begin{pmatrix} 0 \\ 0 \\ -1 \end{pmatrix} = \begin{pmatrix} \partial_1 H_3 - \partial_3 H_1 \\ \partial_2 H_3 - \partial_3 H_2 \\ 0 \end{pmatrix} = \begin{pmatrix} -\partial_3 H_1 \\ -\partial_3 H_2 \\ 0 \end{pmatrix}.$$

Lemma 4.2.2. *Let $1 < q, p < \infty$, $\mu \in \mathbb{R} \setminus \{0\}$, and $f \in \mathcal{P}_\perp L_{q,p}(\mathbb{T} \times \mathbb{R}^3_+)$. Then there exists a unique solution $H \in \mathcal{P}_\perp W^{1,2}_{q,p}(\mathbb{T} \times \mathbb{R}^3_+)$ to (4.12) with $g = 0$ and a constant $c > 0$ such that*

$$\|H\|_{W^{1,2}_{q,p}(\mathbb{T} \times \mathbb{R}^3_+)} \leq c\|f\|_{L_{q,p}(\mathbb{T} \times \mathbb{R}^3_+)}$$

holds. If $f \in \mathcal{P}_\perp L_{q,p,\sigma}(\mathbb{T} \times \mathbb{R}^3_+)$ then the solution H is solenoidal.

Proof: We extend f in the following way: for $i \in \{1, 2\}$ we set

$$\tilde{f}_i = \begin{cases} f_i(x', x_3) & x_3 \geq 0, \\ f_i(x', -x_3) & x_3 < 0, \end{cases}$$

and for the third component we define

$$\tilde{f}_3 = \begin{cases} f_3(x', x_3) & x_3 \geq 0, \\ -f_3(x', -x_3) & x_3 < 0. \end{cases}$$

Since this procedure extends an arbitrary function $g \in C^\infty_{0,\sigma}(\mathbb{T} \times \mathbb{R}^3_+)$ to $\tilde{g} \in C^\infty_{0,\sigma}(G_3)$ we obtain that $\tilde{f} \in \mathcal{P}_\perp L_{q,p,\sigma}(G_3)$ if $f \in \mathcal{P}_\perp L_{q,p,\sigma}(\mathbb{T} \times \mathbb{R}^3_+)$, because $C^\infty_{0,\sigma}(\mathbb{T} \times \mathbb{R}^3_+)$ is dense in $L_{q,p,\sigma}(G_3)$. Lemma 4.2.1 yields a unique solution $H \in \mathcal{P}_\perp W^{1,2}_{q,p}(G_3)$ to (4.12) which is solenoidal if we further assume $f \in \mathcal{P}_\perp L_{q,p,\sigma}(\mathbb{T} \times \mathbb{R}^3_+)$.

Hence, we are left to show that the restriction of H to $\mathbb{T} \times \mathbb{R}^3_+$ satisfies the boundary conditions. For this we define

$$B(t, x) := H_1(t, x', -x_3)e_1 + H_2(t, x', -x_3)e_2 - H_3(t, x', -x_3)e_3.$$

It holds $(\partial_t - \mu\Delta)B = \tilde{f}$ and, since the solution is unique by Lemma 4.2.1, we derive $B = H$. This yields $H_3(t, x', 0) = -H_3(t, x', 0)$ and hence $H \cdot n = H_3 = 0$ on $\mathbb{T} \times \partial\mathbb{R}^3_+$ by (4.14). Furthermore, we conclude $\partial_3 H_i(t, x', 0) = -\partial_3 H_i(t, x', 0)$ for $i \in \{1, 2\}$ and therefore $\partial_3 H_i = 0$ on $\mathbb{T} \times \partial\mathbb{R}^3_+$. Finally, identity (4.15) implies curl $H \times n = 0$, and thus H satisfies the boundary conditions of (4.12) and hence is a solution to this system. Additionally, by (4.13) we derive

$$\|H\|_{W^{1,2}_{q,p}(\mathbb{T}\times\mathbb{R}^3_+)} \leq \|H\|_{W^{1,2}_{q,p}(\mathbb{T}\times\mathbb{R}^3)} \leq c\|\tilde{f}\|_{L_{q,p}(\mathbb{T}\times\mathbb{R}^3)} \leq 2c\|f\|_{L_{q,p}(\mathbb{T}\times\mathbb{R}^3_+)}.$$

To prove uniqueness let u be a solution to (4.12) with $f = g = 0$ and $\varphi \in L_{q',p'}(\mathbb{T} \times \mathbb{R}^3_+)$. By our previous arguments we find a function $v \in W^{1,2}_{q',p'}(\mathbb{T} \times \mathbb{R}^3_+)$ to (4.12) with right hand side φ, $g = 0$ and $-\mu$ instead of μ. Hence, it holds

$$\int_{\mathbb{T}}\int_{\mathbb{R}^3_+} u\varphi \, dx = \int_{\mathbb{T}}\int_{\mathbb{R}^3_+} u(\partial_t v + \mu\Delta v) \, dx = \int_{\mathbb{T}}\int_{\mathbb{R}^3_+} (-\partial_t u + \mu\Delta u)v \, dx = 0$$

by integration by parts. Lemma 1.3.2 finally yields $u = 0$ and therefore uniqueness. $\qquad\square$

For the consideration in Section 4.2.2 we need to show similar estimates for inhomogeneous $g \neq 0$. With the help of the results of Sections 3.3 and 3.4 we have the following.

Lemma 4.2.3. *Let* $1 < q, p < \infty$, $\mu \in \mathbb{R} \setminus \{0\}$,

$$f \in \mathcal{P}_\perp L_{q,p}(\mathbb{T} \times \mathbb{R}^3_+), \quad and \quad g = (g_1, g_2, 0) \in \mathcal{P}_\perp F^{1-\frac{1}{p},(2,1)}_{(q,p),p}(G_2).$$

Then there exists a unique solution $H \in \mathcal{P}_\perp W^{1,2}_{q,p}(\mathbb{T} \times \mathbb{R}^3_+)$ *to (4.12) and a constant* $c > 0$ *such that*

$$\|H\|_{W^{1,2}_{q,p}(\mathbb{T}\times\mathbb{R}^3_+;\mathbb{R}^3)} \leq c\Big(\|f\|_{L_{q,p}(\mathbb{T}\times\mathbb{R}^3_+,\mathbb{R}^3)} + \|g\|_{F^{1-\frac{1}{p},(2,1)}_{(q,p),p}(G_2)}\Big).$$

If $f \in \mathcal{P}_\perp L_{q,p,\sigma}(\mathbb{T} \times \mathbb{R}^3_+)$ *and div* $g = 0$ *on* $\mathbb{R}^2$, *then the solution* H *is solenoidal.*

Proof: From (4.14) and (4.15) we conclude that the boundary condition simplifies to the Neumann boundary condition $\partial_3 H_i = -g_i$ for $i \in \{1, 2\}$ and the Dirichlet boundary condition $H_3 = 0$. Lemma 3.3.3 together with Remark 3.3.4 yields the existence of a purely periodic function $G = (G_1, G_2, 0) \in \mathcal{P}_\perp F^{2,(2,1)}_{(q,p),2}(\mathbb{T} \times \mathbb{R}^3)$ such that G fulfils the boundary conditions in (4.12) and satisfies the estimate

$$(4.16) \qquad\qquad \|G\|_{F^{2,(2,1)}_{(q,p),2}(G_3)} \lesssim \|g\|_{F^{1-\frac{1}{p},(2,1)}_{(q,p),p}(G_2)}.$$

By Corollary 5.2.2 we can apply Lemma 5.2.1 to $h = -\operatorname{div}[G|_{\mathbb{T}\times\mathbb{R}^3_+}]$ and obtain a purely periodic function $\nabla v \in \mathcal{P}_\perp W^{1,2}_{q,p}(\mathbb{T} \times \mathbb{R}^3_+)$ such that $\operatorname{div}(G + \nabla v) = 0$ and $G + \nabla v$ satisfies the boundary conditions in (4.12) since curl $\nabla v = 0$. Proposition 3.1.32 yields $G \in W^{1,2}_{q,p}(G_3)$ and thus

$$(\partial_t - \Delta)[G + \nabla v] \in L_{q,p}(\mathbb{T} \times \mathbb{R}^3_+).$$

Hence, Lemma 4.2.2 yields a solution $\tilde{H}$ to (4.12) with right hand side $f - (\partial_t - \Delta)[G + \nabla v]$ and $g = 0$. Then $H := \tilde{H} + G + \nabla v$ is a solution to (4.12) and we obtain the estimate

$$\|H\|_{\mathrm{W}_{q,p}^{1,2}(\mathbb{T} \times \mathbb{R}_+^3)} \lesssim \|f\|_{\mathrm{L}_{q,p}(\mathbb{T} \times \mathbb{R}_+^3)} + \|G\|_{\mathrm{W}_{q,p}^{1,2}(\mathbb{T} \times \mathbb{R}_+^3)} + \|\nabla v\|_{\mathrm{W}_{q,p}^{1,2}(\mathbb{T} \times \mathbb{R}_+^3)}$$

$$\lesssim \|f\|_{\mathrm{L}_{q,p}(\mathbb{T} \times \mathbb{R}_+^3)} + \|G\|_{\mathrm{W}_{q,p}^{1,2}(G_3)},$$

where we used (5.8) and (5.6) to estimate ∇v by G. Proposition 3.1.24 and (4.16) yield the stated estimate. Uniqueness follows from Lemma 4.2.2, since the difference of two solutions solves (4.12) with $f = g = 0$.

Let $f \in \mathrm{L}_{q,p,\sigma}(\mathbb{T} \times \mathbb{R}_+^3)$ and $\mathrm{div}\, g = 0$ on $\mathbb{R}^2$. Since $\mathrm{div}(G + \nabla v) = 0$ it remains to show

$$(\partial_t - \Delta)[G + \nabla v] \in \mathrm{L}_{q,p,\sigma}(\mathbb{T} \times \mathbb{R}_+^3)$$

to conclude $\mathrm{div}\,\tilde{H} = 0$ by Lemma 4.2.2 and obtain $\mathrm{div}\, H = 0$. The time derivative stays in $\mathrm{L}_{q,p,\sigma}(\mathbb{T} \times \mathbb{R}_+^3)$ in view of (1.10). By the identity of (4.5) we derive

$$\int_{\mathbb{R}_+^3} \Delta[G(t,x) + \nabla v(t,x)] \cdot \nabla \varphi(x)\, \mathrm{d}x = \int_{\mathbb{R}^2} \left[\, \mathrm{curl}\, G(t,y,0) \times \mathrm{e}_3 \,\right] \cdot \nabla \varphi(y,0)\, \mathrm{d}y$$

$$= \int_{\mathbb{R}^2} g(t,y,0) \cdot \nabla \varphi(y,0)\, \mathrm{d}y$$

for any $\varphi \in \mathrm{W}_{p'}^1(\mathbb{R}_+^3)$. The last identity follows from the construction of G. The restriction to the boundary of $\mathbb{R}_+^3$ is an element of $\mathrm{C}_0^\infty(\mathbb{R}^2)$ for every $\varphi \in \mathrm{C}_0^\infty(\overline{\mathbb{R}_+^3})$. Hence the last integral vanishes for all these φ by the properties of g, *i.e.*, $\mathrm{div}\, g = 0$. Since $\mathrm{C}_0^\infty(\overline{\mathbb{R}_+^3})$ is dense in $\mathrm{W}_{p'}^1(\mathbb{R}_+^3)$, we derive that $\Delta[G + \nabla v] \in \mathrm{L}_{q,p,\sigma}(\Omega_\mathbb{T})$ from (1.9). Therefore, $\tilde{H}$ is solenoidal and thus H, which implies the result. $\qquad\square$

Remark 4.2.4. The purely periodicity of f was essential in the previous results to obtain solutions that are elements of $\mathrm{W}_{q,p}^{1,2}$. If the right-hand side f contains a stationary part, then the stationary part of the solution H is only an element of $\widehat{\mathrm{W}}_p^2$ and hence H is no element of $\mathrm{W}_{q,p}^{1,2}$. For details we refer to Kyed and Sauer [61].

4.2.2 Bounded Domains

We have collected enough preparing result to show time-periodic maximal regularity of (4.11) and hence introduce

$$\mathrm{X}_{q,p}^{1,2}(\Omega_\mathbb{T}) := \{H \in \mathrm{W}_{q,p}^{1,2}(\Omega_\mathbb{T}) \mid \mathrm{div}\, H = 0,\ H\big|_{\partial\Omega} \cdot n = 0,\ \mathrm{curl}\, H\big|_{\partial\Omega} \times n = 0\}$$

as the space in which we want to show maximal regularity. The identities of (1.9) and (1.37) yield

(4.17) $$\langle H(t,\cdot) \cdot n, \varphi \rangle = \int_\Omega H(t,x) \cdot \nabla \varphi(x)\, \mathrm{d}x + \int_\Omega \mathrm{div}\, H(t,x)\varphi(x)\, \mathrm{d}x,$$

(4.18) $$\langle \mathrm{curl}\, H(t,\cdot) \times n, \psi \rangle = \int_\Omega \mathrm{curl}\, H(t,x) \cdot \mathrm{curl}\, \psi(x)\, \mathrm{d}x - \int_\Omega \mathrm{curl}\,\mathrm{curl}\, H(t,x) \cdot \psi(x)\, \mathrm{d}x$$

for $\varphi \in \mathrm{W}_{p'}^1(\Omega)$, $\psi \in \mathrm{W}_{p'}^1(\Omega)^3$, $H \in \mathrm{X}_{q,p}^{1,2}(\Omega_\mathbb{T})$ and almost all $t \in \mathbb{T}$. Thus $\mathrm{X}_{q,p}^{1,2}(\Omega_\mathbb{T})$ is a closed subspace of $\mathrm{W}_{q,p}^{1,2}(\Omega_\mathbb{T})$.

The following considerations require a few transformations, hence we state the following result.

Lemma 4.2.5. *Let $k \in \mathbb{N}$, $\varphi : \Omega_1 \to \Omega_2$ be a C^k-diffeomorphism between two domains $\Omega_1, \Omega_2 \subset \mathbb{R}^n$ such that $\|\nabla\varphi\|_{\mathrm{C}^{k-1}(\Omega_1)} + \|\nabla\varphi^{-1}\|_{\mathrm{C}^{k-1}(\Omega_2)} < C$ for some constant $C > 0$. Then $u \mapsto u \circ \varphi$ is a $\mathrm{W}_p^k(\Omega_2) \to \mathrm{W}_p^k(\Omega_1)$ homeomorphism and a $\mathrm{W}_{q,p}^{l,k}(\mathbb{T} \times \Omega_2) \to \mathrm{W}_{q,p}^{l,k}(\mathbb{T} \times \Omega_1)$ homeomorphism for every $l \in \mathbb{N}_0$ and every $1 \le p, q \le \infty$.*

Proof: We start with $p \in [1, \infty)$ and first show that the $\mathrm{W}_p^k(\Omega_1)$-norm of $u \circ \varphi$ can be estimated by the $\mathrm{W}_p^k(\Omega_2)$-norm of u. By the uniform boundedness of φ and integration by substitution it is clear that the $\mathrm{L}_p(\Omega_1)$-norm of $u \circ \varphi$ can the estimated by the $\mathrm{L}_p(\Omega_2)$-norm of u. Let $\alpha \in \mathbb{N}_0^n$ be an arbitrary multi-index with $|\alpha| \le k$. By a multivariate version of higher order of the chain rule and integration by substitution we can estimate the $\mathrm{L}_p(\Omega_1)$-norm of $D^\alpha[u \circ \varphi]$ by the $\mathrm{W}_p^k(\Omega_2)$-norm of u, since there are only finitely many derivatives in the norm of $\mathrm{W}_p^k(\Omega_1)$. This already yields the result, because we can repeat the above steps with φ^{-1} instead of φ. The case of $p = \infty$ follows by the same arguments. The transformation φ does not act on $\mathbb{T}$; therefore, it extends to a homeomorphism $\mathrm{W}_{q,p}^{l,k}(\mathbb{T} \times \Omega_2) \to \mathrm{W}_{q,p}^{l,k}(\mathbb{T} \times \Omega_1)$. $\qquad\square$

Because the following calculations get a bit intricate due to the boundary conditions, we introduce some necessary concepts to ease the understanding. Let $w : \mathbb{R}^2 \to \mathbb{R}$ be a given $\mathrm{C}^1(\mathbb{R}^2)$-function, n denote the normal vector of

$$(4.19) \qquad \mathbb{R}_w^3 := \{x \in \mathbb{R}^3 \mid x_3 > w(x_1, x_2)\}$$

and ν the normal vector of $\mathbb{R}_+^3$. For $\varphi_{-\omega} : \mathbb{R}_\omega^3 \to \mathbb{R}_+^3$ it holds

$$(4.20) \qquad \begin{aligned} \nu \circ \varphi_{-w} &= \frac{\mathrm{cof}(\nabla\varphi_{-w})n}{|\mathrm{cof}(\nabla\varphi_{-w})n|} = \frac{(\nabla\varphi_{-w})^{-\intercal}n}{|(\nabla\varphi_{-w})^{-\intercal}n|} = \frac{(\nabla\varphi_w)^\intercal n}{|(\nabla\varphi_w)^\intercal n|}, \\ \nu &= \frac{(\nabla\varphi_w)^\intercal n \circ \varphi_w}{|(\nabla\varphi_w)^\intercal n \circ \varphi_w|}. \end{aligned}$$

Lemma 4.2.6. *Let $\Omega \subset \mathbb{R}^3$ be a bounded domain of class $\mathrm{C}^{2,1}$ and $1 < q, p < \infty$. Then the operator*

$$(\partial_t - \Delta) : \mathcal{P}_\perp \mathrm{X}_{q,p}^{1,2}(\Omega_\mathbb{T}) \to \mathcal{P}_\perp \mathrm{L}_{q,p,\sigma}(\Omega_\mathbb{T})$$

is injective and has dense range. Furthermore there exists a constant $c > 0$ such that

$$(4.21) \qquad \|H\|_{\mathrm{W}_{q,p}^{1,2}(\Omega_\mathbb{T})} \le c\Big(\|(\partial_t - \Delta)H\|_{\mathrm{L}_{q,p}(\Omega_\mathbb{T})} + \|H\|_{\mathrm{L}_{q,p}(\Omega_\mathbb{T})}\Big)$$

for all $H \in \mathcal{P}_\perp \mathrm{X}_{q,p}^{1,2}(\Omega_\mathbb{T})$.

Proof: We start by proving injectivity and recall that the equations

$$(4.22) \qquad \begin{cases} i\lambda u - \Delta u = f & \text{in } \Omega, \\ \mathrm{div}\, u = 0 & \text{in } \Omega, \\ u \cdot n = 0, \quad \mathrm{curl}\, u \times n = 0 & \text{on } \partial\Omega, \end{cases}$$

admit for every $f \in \mathrm{L}_{p,\sigma}(\Omega)$ and every $\lambda \in \mathbb{R}\backslash\{0\}$ a unique solution in $\mathrm{W}_p^2(\Omega)$ by Theorem 1.8.1. Let $u \in \mathcal{P}_\perp \mathrm{X}_{q,p}^{1,2}(\Omega_\mathbb{T})$ satisfy $(\partial_t - \Delta)u = 0$. Then $[\mathscr{F}_\mathbb{T}u](k, \cdot) \in \mathrm{W}_p^2(\Omega)$ is a solution to equations (4.22) with right hand side 0 and $\lambda = \frac{2\pi}{T}k$ for all $k \in \mathbb{Z} \setminus \{0\}$, and hence $[\mathscr{F}_\mathbb{T}u](k, \cdot) = 0$ because the solution is unique. By u being purely periodic we derive $[\mathscr{F}_\mathbb{T}u](0, \cdot) = 0$ and thus injectivity.

To prove dense range of the operator, we note that Theorem 1.8.1 yields the existence of a solution v to (4.22) with right hand side $f \in L_{p,\sigma}(\Omega)$ and $\lambda = \frac{2\pi}{T} k$ for any $k \in \mathbb{Z} \setminus \{0\}$. Therefore, $u(t,x) = e^{i\frac{2\pi}{T}kt} v(x)$ is an element of $\mathcal{P}_\perp X_{q,p}^{1,2}(\Omega_T)$ and satisfies $(\partial_t - \Delta)u = f e^{i\frac{2\pi}{T}kt}$. The span of this type of functions is dense in $\mathcal{P}_\perp L_{q,p,\sigma}(\Omega_T)$ by Lemma 1.3.4, because applying the projection to the stated dense set of Lemma 1.3.4 yields a dense set of $\mathcal{P}_\perp L_{q,p,\sigma}(\Omega_T)$ by the continuity of $\mathcal{P}_\perp$. Hence we obtain dense range of the operator.

To derive (4.21) we consider an open covering of $\overline{\Omega}$ by $m(\delta)$ balls B_j of radius δ. To this covering we find a partition of unity φ_j such that $\sum_{j=1}^m \varphi(x) = 1$ for all $x \in \overline{\Omega}$ and $\varphi_j \in C_0^\infty(B_j)$. For every $H \in \mathcal{P}_\perp X_{q,p}^{1,2}(\mathbb{T} \times \Omega)$ we define $H_j := \varphi_j H$ and, since $H = \sum_{j=1}^m H_j$, we start by showing estimates for each H_j. If $B_j \subset \Omega$, we can extend H_j by 0 outside of B_j to a function in $W_{q,p}^{1,2}(G_n)$ and hence Lemma 4.2.1 yields the estimate

$$
(4.23) \quad \begin{aligned}
\|H_j\|_{W_{q,p}^{1,2}(\Omega_T)} &\leq c\|(\partial_t - \Delta)H_j\|_{L_{q,p}(\Omega_T)} \\
&\leq c\|\varphi_j(\partial_t - \Delta)H\|_{L_{q,p}(\Omega_T)} + c\|H\Delta\varphi_j - 2\nabla H \cdot \nabla\varphi_j\|_{L_{q,p}(\Omega_T)}.
\end{aligned}
$$

Next we consider $\partial\Omega \cap B_j \neq \emptyset$. By the properties of Ω the boundary part $B_j \cap \partial\Omega$ can be written as a graph of a $C^{2,1}$-function. By rotation and translation we can assume that $(x_1, x_2, \sigma_j(x_1, x_2))$ denotes the boundary part with $\sigma_j \in C^{2,1}(\tilde{B}_j)$ and $\sigma_j(0) = \nabla\sigma_j(0) = 0$. Here $\tilde{B}_j$ is the intersection of B_j with a two-dimensional plane. Without loss of generality, assume that $\tilde{B}_j$ are two-dimensional balls around the origin with radius δ. Lemma 4.1.4 implies that the rotated and translated H remains an element of $\mathcal{P}_\perp X_{q,p}^{1,2}$ and since the operator $(\partial_t - \Delta)$ commutes with rotations and translation and the corresponding norms are equivalent, see Lemma 4.2.5, it suffices to prove an estimate in the stated setting. Theorem 3.3.7 yields an extension of σ_j to ω_j defined on $\mathbb{R}^2$ for any j. Hence we are able to define

$$
\psi_{\omega_j} : \mathbb{R}_+^3 \to \mathbb{R}_{\omega_j}^3 \quad \text{with} \quad \psi_{\omega_j}(x) := (x_1, x_2, x_3 + \omega_j(x_1, x_2)),
$$

where $\mathbb{R}_{\omega_j}^3$ is given by (4.19), and it coincides with $\Omega \cap B_j$ on the support of φ_j. In all these transformations the translations are omitted since they are without any influence. To suit the boundary conditions, we define

$$
\Psi_j H := \nabla\psi_{-\omega_j} H \circ \psi_{\omega_j}.
$$

The functions $\Psi_j H_j$ can be extended by 0 outside of the support of φ_j to functions on $\mathbb{T} \times \mathbb{R}_+^3$ and we denote $H \circ \psi_{\omega_j}$ by $\tilde{H}$ with the same meaning for $\tilde{\varphi}_j$ and $\tilde{H}_j$. To apply Lemma 4.2.3 we need to determine the inhomogeneities f and g. It holds

$$
\begin{aligned}
(\partial_t - \Delta)\Psi_j(H_j) &= \Psi_j(\partial_t H_j) - \Delta(\tilde{\varphi}_j \tilde{H}) + \Delta[(\nabla\omega_j, 0) \cdot \tilde{\varphi}_j \tilde{H}]e_3 \\
&= \Psi_j(\varphi_j \partial_t H) - \tilde{\varphi}_j \Delta\tilde{H} - 2\nabla\tilde{H} \cdot \nabla\tilde{\varphi}_j - \Delta\tilde{\varphi}_j \tilde{H} \\
&\quad + \left[(\nabla\Delta\omega_j, 0) \cdot \tilde{\varphi}_j \tilde{H} + 2\nabla^2\omega_j : \nabla(\tilde{\varphi}_j \tilde{H}) + (\nabla\omega_j, 0) \cdot \Delta(\tilde{\varphi}_j \tilde{H}) \right]e_3 \\
&=: \Psi_j(\varphi_j \partial_t H) - \tilde{\varphi}_j \Delta\tilde{H} - 2\nabla\tilde{H} \cdot \nabla\tilde{\varphi}_j - \Delta\tilde{\varphi}_j \tilde{H} + T(\tilde{\varphi}_j \tilde{H}_j).
\end{aligned}
$$

To keep the expressions somewhat short we set

$$
\Delta\tilde{H} = \left[\Delta H + 2\partial_3 \nabla H \cdot (\nabla\omega_j, 0) + |\nabla\omega_j|^2 \partial_3^2 H + \Delta\omega_j \partial_3 H \right] \circ \psi_{\omega_j} =: (\Delta H) \circ \psi_{\omega_j} + L(H) \circ \psi_{\omega_j},
$$

where ψ_{ω_j} acts on ω_j as the identity. By writing $\nabla\psi_{-\omega_j} =: \text{Id} + A_j$ with $A_j := -e_3 \otimes (\nabla\omega_j, 0)$ we obtain

$$
\begin{aligned}
(\partial_t - \Delta)\Psi_j(H_j) &= \tilde{\varphi}_j(\partial_t H - \Delta H) \circ \psi_{\omega_j} + \tilde{\varphi}_j A_j(\partial_t H) \circ \psi_{\omega_j} - \tilde{\varphi}_j L(H) \circ \psi_{\omega_j} \\
&\quad - 2\nabla\tilde{H} \cdot \nabla\tilde{\varphi}_j - \Delta\tilde{\varphi}_j \tilde{H} + T(\tilde{\varphi}_j \tilde{H}).
\end{aligned}
$$

Next we consider the boundary conditions of $\Psi_j(H_j)$. As stated in (4.20) we have the identity $\nu = (\nabla\psi_{\omega_j})^\mathsf{T} n \circ \psi_{\omega_j} |(\nabla\psi_{\omega_j})^\mathsf{T} n \circ \psi_{\omega_j}|^{-1}$ and therefore

$$\Psi_j(H_j) \cdot \nu = \nabla\psi_{-\omega_j}\varphi_j H \circ \psi_{\omega_j} \cdot \frac{(\nabla\psi_{\omega_j})^\mathsf{T} n \circ \psi_{\omega_j}}{|(\nabla\psi_{\omega_j})^\mathsf{T} n \circ \psi_{\omega_j}|} = \frac{(\varphi_j H \cdot n) \circ \psi_{\omega_j}}{|(\nabla\psi_{\omega_j})^\mathsf{T} n \circ \psi_{\omega_j}|} = 0$$

since $H \cdot n = 0$ and $\nabla\psi_{\omega_j} = (\nabla\psi_{-\omega_j})^{-1}$. For the second boundary condition we derive

$$\operatorname{curl}[\Psi_j(H_j)] = \begin{pmatrix} \partial_2[(-\nabla\omega_j, 1) \cdot (\varphi_j H) \circ \psi_{\omega_j}] - \partial_3[(\varphi_j H_2) \circ \psi_{\omega_j}] \\ \partial_3[(\varphi_j H_1) \circ \psi_{\omega_j}] - \partial_1[(-\nabla\omega_j, 1) \cdot (\varphi_j H) \circ \psi_{\omega_j}] \\ \partial_1[(\varphi_j H_2) \circ \psi_{\omega_j}] - \partial_2[(\varphi_j H_1) \circ \psi_{\omega_j}] \end{pmatrix}$$

$$= \begin{pmatrix} (-\nabla\partial_2\omega_j, 0) \cdot \varphi_j H + (-\nabla\omega_j, 1) \cdot \partial_2(\varphi_j H) + (-\nabla\omega_j, 1)\partial_3(\varphi_j H)\partial_2\omega_j - \partial_3(\varphi_j H_2) \\ \partial_3(\varphi_j H_1) + (\nabla\partial_1\omega_j, 0) \cdot \varphi_j H - (-\nabla\omega_j, 1) \cdot \partial_1(\varphi_j H) - (-\nabla\omega_j, 1) \cdot \partial_3(\varphi_j H)\partial_1\omega_j \\ \partial_1(\varphi_j H_2) + \partial_3(\varphi_j H_2)\partial_1\omega_j - \partial_2(\varphi_j H_1) - \partial_3(\varphi_j H_1)\partial_2\omega_j \end{pmatrix} \circ \psi_{w_j}$$

$$= \operatorname{curl} H_j \circ \psi_{\omega_j} + \begin{pmatrix} (-\nabla\omega_j, 1)\partial_3(\varphi_j H)\partial_2\omega_j - (\nabla\partial_2\omega_j, 0) \cdot \varphi_j H - (\nabla\omega_j, 0) \cdot \partial_2(\varphi_j H) \\ (\nabla\partial_1\omega_j, 0) \cdot \varphi_j H + (\nabla\omega_j, 0) \cdot \partial_1(\varphi_j H) - (-\nabla\omega_j, 1) \cdot \partial_3(\varphi_j H)\partial_1\omega_j \\ \partial_3(\varphi_j H_2)\partial_1\omega_j - \partial_3(\varphi_j H_1)\partial_2\omega_j \end{pmatrix} \circ \psi_{\omega_j}.$$

The normal vector ν to $\mathbb{R}^3_+$ is $-e_3$ and hence

$$\operatorname{curl}[\Psi_j(H_j)] \times \nu = [(\varphi_j \operatorname{curl} H) \circ \psi_{\omega_j}] \times \frac{(\operatorname{Id} - A_j^\mathsf{T})n \circ \psi_{\omega_j}}{|(\operatorname{Id} - A_j^\mathsf{T})n \circ \psi_{\omega_j}|} + (\nabla\varphi_j \times H) \circ \psi_{\omega_j} \times \nu$$

$$+ \begin{pmatrix} (-\nabla\omega_j, 1) \cdot \partial_3(\varphi_j H)\partial_1\omega_j - (\nabla\partial_1\omega_j, 0) \cdot \varphi_j H - (\nabla\omega_j, 0) \cdot \partial_1(\varphi_j H) \\ (-\nabla\omega_j, 1)\partial_3(\varphi_j H)\partial_2\omega_j - (\nabla\partial_2\omega_j, 0) \cdot \varphi_j H - (\nabla\omega_j, 0) \cdot \partial_2(\varphi_j H) \\ 0 \end{pmatrix} \circ \psi_{\omega_j}$$

$$= [(\varphi_j \operatorname{curl} H) \circ \psi_{\omega_j}] \times \frac{e_3[(\nabla\omega_j, 0) \cdot n] \circ \psi_{\omega_j}}{|(\operatorname{Id} - A_j^\mathsf{T})n \circ \psi_{\omega_j}|} + [\nabla\varphi_j \times H] \circ \psi_{\omega_j} \times \nu$$

$$+ \begin{pmatrix} (-\nabla\omega_j, 1) \cdot \partial_3(\varphi_j H)\partial_1\omega_j - (\nabla\partial_1\omega_j, 0) \cdot \varphi_j H - (\nabla\omega_j, 0) \cdot \partial_1(\varphi_j H) \\ (-\nabla\omega_j, 1)\partial_3(\varphi_j H)\partial_2\omega_j - (\nabla\partial_2\omega_j, 0) \cdot \varphi_j H - (\nabla\omega_j, 0) \cdot \partial_2(\varphi_j H) \\ 0 \end{pmatrix} \circ \psi_{\omega_j}.$$

Note that we used $\operatorname{curl} H \times n = 0$ to omit the identity matrix in the first term. To keep the notation in the following estimates somewhat short we recall that $\tilde{H} = H \circ \psi_{\omega_j}$ and shorten $(\nabla H) \circ \psi_{\omega_j}$ to ∇H, meaning that a function ∇H is still composed with the transformation ψ_{ω_j} but no chain rule applies, whereas to a function $\nabla\tilde{H}$ the chain rule applies. With this convention Lemma 4.2.3

yields the estimate

$$
\begin{aligned}
\|\Psi_j H_j\|_{W^{1,2}_{q,p}(\mathbb{T}\times\mathbb{R}^3_+)} &\lesssim \|\varphi_j(\partial_t-\Delta)H\|_{L_{q,p}(\mathbb{T}\times\mathbb{R}^3_+)} + \|\varphi_j A_j \partial_t H\|_{L_{q,p}(\mathbb{T}\times\mathbb{R}^3_+)} + \|\varphi_j L(H)\|_{L_{q,p}(\mathbb{T}\times\mathbb{R}^3_+)} \\
&\quad + 2\|\nabla\tilde{H}\cdot\nabla\tilde{\varphi}_j\|_{L_{q,p}(\mathbb{T}\times\mathbb{R}^3_+)} + \|\Delta\tilde{\varphi}_j\tilde{H}\|_{L_{q,p}(\mathbb{T}\times\mathbb{R}^3_+)} + \|T(\tilde{\varphi}_j\tilde{H})\|_{L_{q,p}(\mathbb{T}\times\mathbb{R}^3_+)} \\
&\quad + \left\|[(\varphi_j\operatorname{curl}H)]\times\frac{e_3[(\nabla\omega_j,0)\cdot n]}{|(\operatorname{Id}-A_j^{\mathsf{T}})n|}\right\|_{F^{1-\frac{1}{p},(2,1)}_{(q,p),p}(G_2)} \\
&\quad + \|\nabla\varphi_j\times H\|_{F^{1-\frac{1}{p},(2,1)}_{(q,p),p}(G_2)} + \|\nabla^2[(\omega_j,0)]\cdot\varphi_j H\|_{F^{1-\frac{1}{p},(2,1)}_{(q,p),p}(G_2)} \\
&\quad + \|\nabla(\varphi_j H)\cdot(\nabla\omega_j,0)\|_{F^{1-\frac{1}{p},(2,1)}_{(q,p),p}(G_2)} + \|(-\nabla\omega_j,1)\cdot\partial_3(\varphi_j H)\partial_1\omega_j\|_{F^{1-\frac{1}{p},(2,1)}_{(q,p),p}(G_2)} \\
&\quad + \|(-\nabla\omega_j,1)\cdot\partial_3(\varphi_j H)\partial_2\omega_j\|_{F^{1-\frac{1}{p},(2,1)}_{(q,p),p}(G_2)} \\
&=: \|D(\varphi_j H)\|_{L_{q,p}(\mathbb{T}\times\mathbb{R}^3_+)} + \|R(\varphi_j H)\|_{F^{1-\frac{1}{p},(2,1)}_{(q,p),p}(G_2)}.
\end{aligned}
$$

From this point on the region of integration of any Lebesgue or Sobolev norm will be the intersection of the half space $\mathbb{T}\times\mathbb{R}^3_+$ with the support of $\tilde{\varphi}_j$ if not stated otherwise. We are going to estimate the term $D(\varphi_j H)$ first. For this we note that φ_j, $\tilde{\varphi}_j$, ω_j and all their occurring derivatives are uniformly bounded. It holds

$$
\begin{aligned}
\text{(4.24)}\qquad \|D(\varphi_j H)\|_{L_{q,p}(\mathbb{T}\times\mathbb{R}^3_+)} &\lesssim \|(\partial_t-\Delta)H\|_{L_{q,p}} + \|\nabla\tilde{H}\|_{L_{q,p}} + \|\tilde{H}\|_{L_{q,p}} \\
&\quad + \|\nabla\omega_j\|_{L_\infty}\left(\|\partial_t\tilde{H}_j\|_{L_{q,p}} + \|\Delta\tilde{H}_j\|_{L_{q,p}} + \|\varphi_j\partial_3\nabla H\|_{L_{q,p}}\right) \\
&\quad + \|\nabla\omega_j\|^2_{L_\infty}\|\varphi_j\partial_3^2 H\|_{L_{q,p}}.
\end{aligned}
$$

Using the identities $\partial_3\nabla(\varphi_j H) = \partial_3\varphi_j\nabla H + \varphi_j\partial_3\nabla H + \partial_3 H\otimes\nabla\varphi_j + H\otimes\partial_3\nabla\varphi_j$ and $\partial_3^2(\varphi_j H) = \partial_3^2\varphi_3 H + 2\partial_3\varphi_j\partial_3 H + \varphi_j\partial_3^2 H$, absorbing the terms of lower order of derivatives of H into the first line of estimate of the right-hand side of (4.24) by Lemma 4.2.5 and applying said lemma to only get terms of $\tilde{H}$ or $\tilde{H}_j$, we obtain

$$
\text{(4.25)}\qquad \|D(\varphi_j H)\|_{L_{q,p}(\mathbb{T}\times\mathbb{R}^3_+)} \lesssim \|(\partial_t-\Delta)H\|_{L_{q,p}} + \|\tilde{H}\|_{W^{0,1}_{q,p}} + \|\nabla\omega_j\|_{L_\infty}\|\tilde{H}_j\|_{W^{1,2}_{q,p}}.
$$

Next consider $R(\varphi_j H)$ and recall Corollary 3.2.9, allowing us to estimate products in the Triebel-Lizorkin norm, and Corollary 3.4.7, which grants estimates of the trace, to derive

$$
\begin{aligned}
\|R(\varphi_j H)\|_{F^{1-\frac{1}{p},(2,1)}_{(q,p),p}(G_2)} &\lesssim \|\nabla\varphi_j\times H\|_{F^{1,(2,1)}_{(q,p),2}(\mathbb{T}\times\mathbb{R}^3_+)} + \|\nabla^2\omega_j\|_{B^{1-\frac{1}{2p},(2,1)}_{\infty,\infty}(G_2)}\|\varphi_j H\|_{F^{1,(2,1)}_{(q,p),2}(\mathbb{T}\times\mathbb{R}^3_+)} \\
&\quad + \|\nabla\omega_j\|_{B^{1-\frac{1}{2p},(2,1)}_{\infty,\infty}(G_2)}\|[\nabla(\varphi_j H)]\circ\psi_{\omega_j}\|_{F^{1,(2,1)}_{(q,p),2}(\mathbb{T}\times\mathbb{R}^3_+)} \\
&\quad + \|[\varphi_j\operatorname{curl}H]\circ\psi_{\omega_j}\|_{F^{1,(2,1)}_{(q,p),2}(\mathbb{T}\times\mathbb{R}^3_+)}\left\|\frac{[(\nabla\omega_j,0)\cdot n]}{|(\operatorname{Id}-A_j^{\mathsf{T}})n|}\right\|_{B^{1-\frac{1}{2p},(2,1)}_{\infty,\infty}(G_2)} \\
&\quad + \|\nabla\omega_j\|_{B^{1-\frac{1}{2p},(2,1)}_{\infty,\infty}(G_2)}\|(\nabla\omega_j,1)\|_{B^{1-\frac{1}{2p},(2,1)}_{\infty,\infty}(G_2)}\|\varphi_j H\|_{F^{2,(2,1)}_{(q,p),2}(\mathbb{T}\times\mathbb{R}^3_+)}.
\end{aligned}
$$

Note that ∂_3 commutes with the transformation ψ_{ω_j} hence we directly applied Lemma 3.1.33. Furthermore, we chose $s_1 = 1-\frac{1}{2p} > 1-\frac{1}{p}$ in the application of Corollary 3.2.9. Lemma 3.3.10 implies that the Besov norm of $(\nabla\omega_j,1)$ and $\nabla\psi_{-\omega_j}$ are uniformly bounded since we can, without loss of generality, assume that the radius of the balls satisfies $\delta < 1$. Furthermore, we have $[\nabla(\varphi_j H)] \circ \psi_{\omega_j} = \nabla(\tilde{\varphi}_j \tilde{H}) \cdot \nabla \psi_{-\omega_j}$, hence applications of Lemma 3.1.34, Lemma 3.3.9 and Lemma 3.1.33 yield

$$
\begin{aligned}
\|R(\varphi_j H)\|_{F_{(q,p),p}^{1-\frac{1}{p},(2,1)}(G_2)} &\lesssim \|H\|_{L_{q,p}}^{\frac{1}{2}} \|H\|_{W_{q,p}^{1,2}}^{\frac{1}{2}} + \|\nabla^2 \omega_j\|_{C^{0,1-\frac{1}{2p}}(G_2)} \|H\|_{L_{q,p}}^{\frac{1}{2}} \|H\|_{W_{q,p}^{1,2}}^{\frac{1}{2}} \\
&\quad + \|\nabla \omega_j\|_{C^{0,1-\frac{1}{2p}}(G_2)} \|\varphi_j H\|_{F_{(q,p),p}^{2,(2,1)}(\mathbb{T} \times \mathbb{R}_+^3)} \\
&\quad + \|\nabla(\varphi_j H)\|_{F_{(q,p),2}^{1,(2,1)}(\mathbb{T} \times \mathbb{R}_+^3)} \left\| \frac{[(\nabla \omega_j, 0) \cdot n]}{|(\mathrm{Id} - A_j^\intercal)n|} \right\|_{C^{0,1-\frac{1}{2p}}(G_2)} \\
&\quad + \|\nabla \varphi_j H\|_{F_{(q,p),2}^{1,(2,1)}(\mathbb{T} \times \mathbb{R}_+^3)} \left\| \frac{[(\nabla \omega_j, 0) \cdot n]}{|(\mathrm{Id} - A_j^\intercal)n|} \right\|_{C^{0,1-\frac{1}{2p}}(G_2)}.
\end{aligned}
$$
(4.26)

Before we continue we remark that we can exchange G_2 by $\mathbb{R}^2$ in the norms of ω_j since the functions do not depend on time, and because it is clear that ω_j is defined on $\mathbb{R}^2$ and σ_j on B_δ, we will omit the domain in the Hölder norms. It holds

$$
\begin{aligned}
\left\| \frac{[(\nabla \omega_j, 0) \cdot n]}{|(\mathrm{Id} - A_j^\intercal)n|} \right\|_{C^{0,1-\frac{1}{2p}}(G_2)} &\leq \|\nabla \omega_j\|_{C^0} \||(\mathrm{Id} - A_j^\intercal)|^{-1}\|_{C^0} + \||(\mathrm{Id} - A_j^\intercal)|^{-1}\|_{C^0} \|(\nabla \omega_j, 0) \cdot n\|_{\hat{C}^{1-\frac{1}{2p}}} \\
&\quad + \|\nabla \omega_j\|_{C^0} \||(\mathrm{Id} - A_j^\intercal)|^{-1}\|_{C^0}^2 \|(\mathrm{Id} - A_j^\intercal)n\|_{\hat{C}^{1-\frac{1}{2p}}}.
\end{aligned}
$$

Recall that the matrix A_j has the form $-e_3 \otimes (\nabla \omega_j, 0)$. Theorem 3.3.7 yields $\|\nabla \omega_j\|_{C^0(\mathbb{R}^2)} \leq C_1 \|\sigma_j\|_{C^1(B_\delta)}$. Since the compact boundary is of class $C^{2,1}$, we can choose δ small enough, such that we have $C_1 \|\sigma_j\|_{C^1(B_\delta)} < \frac{1}{2}$ for all j as $\sigma_j(0) = \nabla \sigma_j(0) = 0$. This implies $|(\mathrm{Id} - A_j^\intercal)|^{-1} \leq 2$ and because the normal vector of the boundary has a uniformly bounded $C^{1,1}$-norm, we conclude with Theorem 3.3.7 and Lemma 3.3.10

$$
\begin{aligned}
\left\| \frac{[(\nabla \omega_j, 0) \cdot n]}{|(\mathrm{Id} - A_j^\intercal)n|} \right\|_{C^{0,1-\frac{1}{2p}}} &\lesssim \|\sigma_j\|_{C^1} + \|\nabla \omega_j\|_{C^{0,1-\frac{1}{2p}}} + \|\sigma_j\|_{C^1} \|(\mathrm{Id} - A_j^\intercal)\|_{C^{0,1-\frac{1}{2p}}} \\
&\lesssim \|\sigma_j\|_{C^1} + \|\sigma_j\|_{C^1} + \delta^{\frac{1}{2p}} \|\sigma_j\|_{C^2} + \|\sigma_j\|_{C^1} \left[1 + \|\nabla \omega_j\|_{C^{0,1-\frac{1}{2p}}}\right] \\
&\lesssim \|\sigma_j\|_{C^1} \left[1 + \|\sigma_j\|_{C^1} + \delta^{\frac{1}{2p}} \|\sigma_j\|_{C^2}\right] + \delta^{\frac{1}{2p}} \|\sigma_j\|_{C^2}.
\end{aligned}
$$

Since the boundary is compact, the $C^{2,1}$-norm of σ_j is uniformly bounded with respect to j and δ, and for $\delta < 1$ it holds

$$
\left\| \frac{[(\nabla \omega_j, 0) \cdot n]}{|(\mathrm{Id} - A_j^\intercal)n|} \right\|_{C^{0,1-\frac{1}{2p}}} \lesssim \|\sigma_j\|_{C^1} + \delta^{\frac{1}{2p}} \|\sigma_j\|_{C^2}.
$$

Repeating the arguments used to obtain (4.26) together with Theorem 3.3.7 and Lemma 3.3.10 we conclude

$$
\begin{aligned}
\|R(\varphi_j H)\|_{F_{(q,p),p}^{1-\frac{1}{p},(2,1)}(G_2)} &\leq C(\delta) \|H\|_{L_{q,p}}^{\frac{1}{2}} \|H\|_{W_{q,p}^{1,2}}^{\frac{1}{2}} \\
&\quad + C\left(\|\sigma_j\|_{C^1} + \delta^{\frac{1}{2p}} \|\sigma_j\|_{C^2}\right) \|\varphi_j H\|_{W_{q,p}^{1,2}}
\end{aligned}
$$
(4.27)

with a constant $C > 0$ independent of δ. By combining the estimates of (4.25) and (4.27), we derive

$$
\begin{aligned}
\|\Psi_j H_j\|_{W_{q,p}^{1,2}(\mathbb{T} \times \mathbb{R}_+^3)} &\leq C \|(\partial_t - \Delta)H\|_{L_{q,p}} + C \|\tilde{H}\|_{W_{q,p}^{0,1}} + C \|\sigma_j\|_{C^1} \|\tilde{H}_j\|_{W_{q,p}^{1,2}} \\
&\quad + C(\delta) \|H\|_{L_{q,p}}^{\frac{1}{2}} \|H\|_{W_{q,p}^{1,2}}^{\frac{1}{2}} + C\left(\|\sigma_j\|_{C^1} + \delta^{\frac{1}{2p}} \|\sigma_j\|_{C^2}\right) \|\varphi_j H\|_{W_{q,p}^{1,2}}.
\end{aligned}
$$

Since the matrix $\nabla\psi_{-\omega_j}$ is invertible with $\left(\nabla\psi_{-\omega_j}\right)^{-1} = \nabla\psi_{\omega_j}$ and both are uniformly bounded, we have that the $W^{1,2}_{q,p}$-norms of $\nabla\psi_{-\omega_j}H$ and H are equivalent. Next we choose δ sufficiently small so that the $W^{1,2}_{q,p}$-norm of H_j on the right hand side can be absorbed by the expression on the left-hand side. Note that $\|\sigma_j\|_{C^1}$ gets smaller with smaller δ since $\sigma_j(0) = \nabla\sigma_j(0) = 0$. This yields the estimate

$$(4.28) \qquad \|H_j\|_{W^{1,2}_{q,p}} \lesssim \|(\partial_t - \Delta)H\|_{L_{q,p}} + \|H\|_{W^{0,1}_{q,p}} + \|H\|^{\frac{1}{2}}_{L_{q,p}}\|\tilde{H}\|^{\frac{1}{2}}_{W^{1,2}_{q,p}}.$$

Here we applied Lemma 4.2.5 so no transformations occur. Since δ is now fixed, we obtain finitely many φ_j. By summing up over all of them and using $\sum \varphi_j H = H$ together with the estimates (4.23) and (4.28), we obtain a constant $K > 0$ such that

$$(4.29) \qquad \|H\|_{W^{1,2}_{q,p}} \leq K\|(\partial_t - \Delta)H\|_{L_{q,p}} + K\|H\|_{W^{0,1}_{q,p}} + K\|H\|^{\frac{1}{2}}_{L_{q,p}}\|H\|^{\frac{1}{2}}_{W^{1,2}_{q,p}}.$$

Now we apply the standard inequality

$$(4.30) \qquad \|u\|_{W^1_p(\Omega)} \leq C(\varepsilon)\|u\|_{L_p(\Omega)} + \varepsilon\|u\|_{W^2_p(\Omega)},$$

see Adams and Fournier [1, Lemma 5.2] for example, to the second term on the right-hand side of (4.29) and Young's inequality to the third one in a way that the factor in front of $\|H\|_{W^{1,2}_{q,p}}$ and $\|H\|_{W^{0,2}_{q,p}}$ is $\frac{1}{4K}$. Hence, the estimate (4.21) follows by absorbing the higher order terms. $\qquad\square$

As a next step we will show that in (4.21) the extra term $\|H\|_{L_{q,p}}(\Omega_{\mathbb{T}})$ can be omitted by a compactness argument.

Theorem 4.2.7. *For every $1 < q, p < \infty$ and every $f \in \mathcal{P}_\perp L_{q,p,\sigma}(\Omega_{\mathbb{T}})$ we get a unique solution H to (4.11) in $\mathcal{P}_\perp X^{1,2}_{q,p}(\Omega_{\mathbb{T}})$ and a constant $c(\mu,\Omega)$ such that*

$$(4.31) \qquad \|H\|_{W^{1,2}_{q,p}(\Omega_{\mathbb{T}})} \leq c\|(\partial_t - \mu\Delta)H\|_{L_{q,p}(\Omega_{\mathbb{T}})} = \|f\|_{L_{q,p}(\Omega_{\mathbb{T}})}.$$

Proof: Let us first assume that (4.31) does not hold. Therefore, we find a sequence of functions $\{H_k\}_{k\in\mathbb{N}} \subset \mathcal{P}_\perp X^{1,2}_{q,p}(\Omega_{\mathbb{T}})$ such that $\|H_k\|_{W^{1,2}_{q,p}} = 1$ and $\|(\partial_t - \Delta)H_k\|_{L_{q,p}} \to 0$ for $k \to \infty$. By Lemma 5.1.8 we get a subsequence $\{H_{k_n}\}_{n\in\mathbb{N}}$ that converges weakly to some H in $W^{1,2}_{q,p}(\Omega_{\mathbb{T}})$. The weak convergence implies convergence of the integrals in the equations (4.17) and (4.18), hence the limit H satisfies the boundary conditions. As weak convergence implies convergence in $\mathcal{D}'(\Omega_{\mathbb{T}})$ the function H is purely periodic. Furthermore H satisfies (4.11) with $f = 0$, since the right hand side converges to 0 by our assumptions. Therefore, we obtain $H = 0$ by Lemma 4.2.6. The compactness result of Lemma 5.1.8 yields $\|H_{k_n}\|_{L_{q,p}} \to 0$ for $n \to \infty$. This results in a contradiction by Lemma 4.2.6 since

$$1 = \lim_{n\to\infty} \|H_{k_n}\|_{W^{1,2}_{q,p}(\mathbb{T}\times\Omega)} \lesssim \lim_{n\to\infty} \left(\|(\partial_t - \Delta)H_{k_n}\|_{L_{q,p}(\Omega_{\mathbb{T}})} + \|H_{k_n}\|_{L_{q,p}(\Omega_{\mathbb{T}})}\right) = 0.$$

Hence, the operator

$$(\partial_t - \Delta) : \mathcal{P}_\perp X^{1,2}_{q,p}(\Omega_{\mathbb{T}}) \to \mathcal{P}_\perp L_{q,p,\sigma}(\Omega_{\mathbb{T}})$$

has a closed range and by Lemma 4.2.6 it is a homeomorphism. Consequently, the operator is invertible and we obtain a unique solution to every $f \in \mathcal{P}_\perp L_{q,p,\sigma}(\Omega_{\mathbb{T}})$ which satisfies (4.31). $\qquad\square$

4.3 Perturbed Linear Theory

Now that we have shown important estimates for the unperturbed system, we will consider the perturbed linear system and construct unique time-periodic solutions with suitable estimates. The system is given by

(4.32)
$$\begin{cases} \partial_t u - \Delta u - (H_0 \cdot \nabla)H - (H \cdot \nabla)H_0 + \nabla \mathfrak{p} = F & \text{in } \Omega_{\mathbb{T}}, \\ \partial_t H - \Delta H - \nabla \times [u \times H_0] = G & \text{in } \Omega_{\mathbb{T}}, \\ \operatorname{div} u = \operatorname{div} H = 0 & \text{in } \Omega_{\mathbb{T}}, \\ u = 0, \quad H \cdot n = 0, \quad \operatorname{curl} H \times n = 0 & \text{on } \partial\Omega_{\mathbb{T}}. \end{cases}$$

Here $\Omega \subset \mathbb{R}^3$ is a bounded simply connected domain of class $C^{2,1}$ and the functions $F \in L_{q,p}(\Omega_{\mathbb{T}})$ and $G \in L_{q,p,\sigma}(\Omega_{\mathbb{T}})$ are given. Since we are directly considering the problem on a bounded domain and do not need to apply any localization methods as in the proof of Theorem 4.2.6, we do not need to split the problem into a purely periodic and a stationary part. In order to show existence of a solution to (4.32), some ideas of the proofs of Lemma 4.2.6 and Theorem 4.2.7 are needed. Hence, recall

$$X_{q,p}^{1,2}(\Omega_{\mathbb{T}}) := \{H \in W_{q,p}^{1,2}(\Omega_{\mathbb{T}}) \mid \operatorname{div} H = 0, \, H\big|_{\partial\Omega} \cdot n = 0, \, \operatorname{curl} H\big|_{\partial\Omega} \times n = 0\}$$

and introduce

$$Y_{q,p}^{1,2}(\Omega_{\mathbb{T}}) := \{u \in W_{q,p}^{1,2}(\Omega_{\mathbb{T}}) \mid \operatorname{div} u = 0, \, u\big|_{\partial\Omega} = 0\}.$$

Theorem 4.3.1. *Let* $1 < q, p < \infty$, $(F, G) \in L_{q,p}(\Omega_{\mathbb{T}}) \times L_{q,p,\sigma}(\Omega_{\mathbb{T}})$ *and* $H_0 \in W_\infty^1(\Omega)$ *with* $\nabla H_0 = (\nabla H_0)^\intercal$. *Then there exists a unique solutions* $(u, H, p) \in W_{q,p}^{1,2}(\Omega_{\mathbb{T}})^2 \times \widehat{W}_{q,p}^{0,1}(\Omega_{\mathbb{T}})$ *to (4.32) such that*

(4.33)
$$\|u\|_{W_{q,p}^{1,2}(\Omega_{\mathbb{T}})} + \|H\|_{W_{q,p}^{1,2}(\Omega_{\mathbb{T}})} + \|\mathfrak{p}\|_{W_{q,p}^{0,1}(\Omega_{\mathbb{T}})} \le c(\kappa)\Big(\|F\|_{L_{q,p}(\Omega_{\mathbb{T}})} + \|G\|_{L_{q,p}(\Omega_{\mathbb{T}})}\Big),$$

where the constant $c(\kappa)$ *can be chosen independent of* H_0 *if* $\|H_0\|_{W_\infty^1(\Omega)} < \kappa$ *is satisfied for some constant* $\kappa > 0$.

Proof: We introduce the operator

$$Z_p(u, H, \mathfrak{p}) : Y_{q,p}^{1,2}(\Omega_{\mathbb{T}}) \times X_{q,p}^{1,2}(\Omega_{\mathbb{T}}) \times \widehat{W}_{q,p}^{0,1}(\Omega_{\mathbb{T}}) \to L_{q,p}(\Omega_{\mathbb{T}}) \times L_{q,p,\sigma}(\Omega_{\mathbb{T}}),$$
$$Z_p(u, H, \mathfrak{p}) := (\partial_t - \Delta)(u, H) + (\nabla \mathfrak{p}, 0) + (S_p^1(H), S_p^2(u)),$$

with S_p^1 and S_p^2 as in (4.8), and show that it is a homeomorphism. We start by showing that Z_p is injective and assume $Z_p(v, B, \mathfrak{p}) = 0$ for $(v, B, \mathfrak{p}) \in Y_{q,p}^{1,2}(\Omega_{\mathbb{T}}) \times X_{q,p}^{1,2}(\Omega_{\mathbb{T}}) \times \widehat{W}_{q,p}^{0,1}(\Omega_{\mathbb{T}})$. By applying the Helmholtz projection $\mathcal{P}_H$ we see that

$$i\frac{2\pi}{T}k\big(\mathscr{F}_{\mathbb{T}}[v](k, \cdot), \mathscr{F}_{\mathbb{T}}[B](k, \cdot)\big) + T_p(\mathscr{F}_{\mathbb{T}}[v](k, \cdot), \mathscr{F}_{\mathbb{T}}[B](k, \cdot)) = 0$$

for every $k \in \mathbb{Z}$, with T_p given as in (4.9). Hence, by Theorem 4.1.7 we derive $\mathscr{F}_{\mathbb{T}}[v](k, \cdot) = \mathscr{F}_{\mathbb{T}}[B](k, \cdot) = 0$ for every $k \in \mathbb{Z}$ and, by the properties of the Fourier transform, $v = B = 0$. Thus, $\nabla \mathfrak{p} = 0$ since all other terms of (4.32) are zero, and therefore injectivity follows because we are considering the pressure to be in $\widehat{W}_{q,p}^{0,1}(\Omega_{\mathbb{T}})$.

As a next step we will show that the operator has dense range. Lemma 1.3.4 implies that the span of $\varphi_k(t) := e^{i\frac{2\pi}{T}kt}h$ is dense in $L_{q,p}(\Omega_T)$ if $h \in L_p(\Omega)$ and dense in $L_{q,p,\sigma}(\Omega_T)$ if $h \in L_{p,\sigma}(\Omega)$. Therefore, it suffices to construct a solution of (4.32) with right-hand side $(e^{i\frac{2\pi}{T}lt}f, e^{i\frac{2\pi}{T}mt}g)$ for every $l, m \in \mathbb{Z}$, $f \in L_p(\Omega)$ and $g \in L_{p,\sigma}(\Omega)$. By the results of Theorem 4.1.7 we find solutions (u_l, H_l) and (u_m, H_m) such that $i\frac{2\pi}{T}l(u_l, H_l) + T_p(u_l, H_l) = (\mathcal{P}_H f, 0)$ and $i\frac{2\pi}{T}m(u_m, H_m) + T_p(u_m, H_m) = (0, g)$ with $u_l, u_m \in D(A_p)$ and $H_l, H_m \in D(B_p)$. By the properties of the Helmholtz projection, see Lemma 1.3.8, we obtain corresponding pressure terms $\mathfrak{p}_l$ and $\mathfrak{p}_m$ such that

$$Z_p\left(e^{i\frac{2\pi}{T}lt}u_l + e^{i\frac{2\pi}{T}mt}u_m, e^{i\frac{2\pi}{T}lt}H_l + e^{i\frac{2\pi}{T}mt}H_m, e^{i\frac{2\pi}{T}lt}\mathfrak{p}_l + e^{i\frac{2\pi}{T}mt}\mathfrak{p}_m\right) = (e^{i\frac{2\pi}{T}lt}f, e^{i\frac{2\pi}{T}mt}g).$$

Since f, g, m, l were arbitrary, we derive that the operator Z_p has dense range.

To prove that the range of Z_p is closed we consider $(u, H, \mathfrak{p}) \in Y_{q,p}^{1,2}(\Omega_T) \times X_{q,p}^{1,2}(\Omega_T) \times \widehat{W}_{q,p}^{0,1}(\Omega_T)$ and define $u_s := \mathcal{P}u$ and $u_\perp := \mathcal{P}_\perp u$ with similar definitions for $H_s, H_\perp$ and $\nabla \mathfrak{p}_s, \nabla \mathfrak{p}_\perp$. Then $u_\perp$ solves

$$\begin{cases} \partial_t u_\perp - \Delta u_\perp + \nabla \mathfrak{p}_\perp = \left(Z_p(u_\perp, H_\perp, \mathfrak{p}_\perp)\right)_1 - S_p^1(H_\perp) & \text{in } \Omega_T, \\ \operatorname{div} u = 0 & \text{in } \Omega_T, \\ u|_{\partial\Omega} = 0, & \text{on } \partial\Omega_T. \end{cases}$$

For the last part it is important that H_0 is independent of time, since otherwise the projections would not commute with the operators S_p^1 and S_p^2. Since the right-hand side is in $\mathcal{P}_\perp L_{q,p}(\Omega_T)$, we obtain by Maekawa and Sauer [66, Theorem 4.11] the estimate

$$\|u_\perp\|_{W_{q,p}^{1,2}(\Omega_T)} + \|\nabla \mathfrak{p}_\perp\|_{L_{q,p}(\Omega_T)} \lesssim \|Z_p(u, H, \mathfrak{p})\|_{L_{q,p}(\Omega_T)} + \|S_p^1(H)\|_{L_{q,p}(\Omega_T)}.$$

Here we additionally estimated the terms on the right-hand side by the continuity of $\mathcal{P}_\perp$, because it commutes with derivatives. The stationary part u_s solves the same equation without the time derivative and therefore Galdi [35, Lemma IV.6.1] implies the estimate

$$\|u_s\|_{W_{q,p}^{1,2}(\Omega_T)} + \|\nabla \mathfrak{p}_s\|_{L_{q,p}(\Omega_T)} \lesssim \|Z_p(u, H, \mathfrak{p})\|_{L_{q,p}(\Omega_T)} + \|S_p^1(H)\|_{L_{q,p}(\Omega_T)}.$$

To estimate $H_\perp$ and H_s we employ the same idea and use Theorem 4.2.7 and Proposition 1.8.2 instead of the result of Maekawa and Sauer or Galdi, respectively, to obtain

$$\|(u, H)\|_{W_{q,p}^{1,2}(\Omega_T)} + \|\mathfrak{p}\|_{\widehat{W}_{q,p}^{0,1}(\Omega_T)} \lesssim \|Z_p(u, H, \mathfrak{p})\|_{L_{q,p}(\Omega_T)} + \|(S_p^1(H), S_p^2(u))\|_{L_{q,p}(\Omega_T)}.$$

Similar to (4.10) we obtain

$$\|(S_p^1(H), S_p^2(u))\|_{L_p(\Omega_T)} \lesssim c(H_0)\|(u, H)\|_{W^{0,1}(\Omega_T)} \leq C(\varepsilon, H_0)\|(u, H)\|_{L_{q,p}(\Omega_T)} + \varepsilon\|(u, H)\|_{W_{q,p}^{1,2}(\Omega_T)}$$

by using (4.30) instead of Lemma 4.1.5. Choosing ε suitably small and absorbing the highest order terms by the left-hand side yields the estimate

$$(4.34) \qquad \|(u, H)\|_{W_{q,p}^{1,2}(\Omega_T)} + \|\mathfrak{p}\|_{\widehat{W}_{q,p}^{0,1}(\Omega_T)} \leq c\|Z_p(u, H, \mathfrak{p})\|_{L_{q,p}(\Omega_T)} + c(H_0)\|(u, H)\|_{L_{q,p}(\Omega_T)}.$$

We want to show that inequality (4.34) holds without the lower order term on the right-hand side and a constant depending on H_0 as stated in the theorem. For this, we assume that the inequality does not hold in the stated form. Hence we find a sequence $\{(u_n, H_n, \mathfrak{p}_n)\}_{n \in \mathbb{N}}$ and a sequence $\{H_0^n\}_{n \in \mathbb{N}}$ such that

$$\|(u_n, H_n)\|_{W_{q,p}^{1,2}(\Omega_T)} + \|\mathfrak{p}_n\|_{\widehat{W}_{q,p}^{0,1}(\Omega_T)} = 1, \quad \|H_0^n\|_{W_\infty^1(\Omega)} \leq \kappa \quad \text{and}$$

$$\|Z_p(u_n, H_n, \mathfrak{p}_n)\|_{L_{q,p}(\Omega_T)} \to 0 \text{ for } n \to \infty.$$

Without loss of generality we can assume that $\mathfrak{p}_n$ has mean value 0 and hence $\mathfrak{p}_n \in W_{q,p}^{0,1}(\Omega_T)$ by Poincaré's inequality. Therefore, by Lemma 5.1.8 we find a subsequence $\{(u_{n_k}, H_{n_k}, \mathfrak{p}_{n_k})\}_{k \in \mathbb{N}}$ that converges weakly to some $(u, H, \mathfrak{p})$. Because $L_1(\Omega)$ is separable, we obtain a subsequence $\{H_0^{n_k}\}_{k \in \mathbb{N}}$ and a $H_0 \in W_\infty^1(\Omega)$ such that $H_0^{n_k}$ converges weakly-* to H_0 in $L_\infty(\Omega)$ and $\{\nabla H_0^{n_k}\}_{k \in \mathbb{N}}$ converges weakly-* to ∇H_0 in L_∞. Furthermore, H_0 satisfies $\nabla H_0 = (\nabla H_0)^\intercal$ as each $H_0^{n_k}$ does. For any $\varphi \in C_0^\infty(\Omega_T)$ we have

$$\langle H_0^{n_k} u_{n_k} - H_0 u, \varphi \rangle = \langle H_0^{n_k} u_{n_k} - H_0^{n_k} u, \varphi \rangle + \langle H_0^{n_k} - H_0, u\varphi \rangle \to 0 \quad \text{for } k \to \infty$$

since u_{n_k} converges strongly to u in $W_{q,p}^{0,1}(\Omega_T)$ by Lemma 5.1.8, $H_0^{n_k}$ is uniformly bounded in $L_\infty(\Omega)$, and $u\varphi \in L_1(\Omega_T)$. Repeating this argument implies that all terms in S_p^1 and S_p^2 converge weakly in the same way. Because $(\partial_t - \Delta)(u, H)$ converges weakly, we obtain that $(u, H, \mathfrak{p})$ satisfies (4.32) with right-hand side $F = G = 0$. Since the operator is injective, we derive $(u, H, \mathfrak{p}) = (0, 0, 0)$ and the compactness of Lemma 5.1.8 implies $\|(u_{n_k}, H_{n_k})\|_{L_{q,p}(\Omega_T)} \to 0$ for $k \to \infty$. This results in a contradiction with (4.34) because

$$1 = \lim_{k \to \infty} \|(u_{n_k}, H_{n_k})\|_{W_{q,p}^{1,2}(\Omega_T)} + \|\mathfrak{p}_{n_k}\|_{\widehat{W}_{q,p}^{0,1}(\Omega_T)}$$

$$\leq \lim_{k \to \infty} \left[\|Z_p(u_{n_k}, H_{n_k}, \mathfrak{p}_{n_k})\|_{L_{q,p}(\Omega_T)} + c(H_0)\|(u_{n_k}, H_{n_k})\|_{L_{q,p}(\Omega_T)} \right] = 0.$$

Hence, we derive the estimate

(4.35) $$\|(u, H)\|_{W_{q,p}^{1,2}(\Omega_T)} + \|\mathfrak{p}\|_{\widehat{W}_{q,p}^{0,1}(\Omega_T)} \leq c(\kappa)\|Z_p(u, H, \mathfrak{p})\|_{L_{q,p}(\Omega_T)}$$

for any $H_0 \in W_\infty^1(\Omega)$ such that $\|H_0\|_{W_\infty^1(\Omega)} \leq \kappa$ and $\nabla H_0 = (\nabla H_0)^\intercal$. This further implies that the operator has closed range and thus is a homeomorphism. Therefore, we obtain a unique solution of (4.32) for every $F \in L_{q,p}(\Omega_T)$ and $G \in L_{q,p,\sigma}(\Omega_T)$. By Poincaré's inequality and (4.35) we conclude

$$\|(u, H)\|_{W_{q,p}^{1,2}(\Omega_T)} + \|\mathfrak{p}\|_{W_{q,p}^{0,1}(\Omega_T)} \leq c(\kappa)\|(F, G)\|_{L_{q,p}(\mathbb{T} \times \Omega)}$$

for any $H_0 \in W_\infty^1(\Omega)$ with the stated properties. $\qquad \square$

4.4 The Nonlinear Problem

We recall the non-linear problem after the transformation with the extended boundary data given by

(4.36)
$$\begin{cases} \partial_t u - \Delta u - S_p^1(H) + \nabla\mathfrak{p} = F(H_0) + (H \cdot \nabla)H - (u \cdot \nabla)u & \text{in } \Omega_T, \\ \partial_t H - \Delta H - \nabla \times [u \times H_0] = \nabla \times [u \times H] & \text{in } \Omega_T, \\ \operatorname{div} u = \operatorname{div} H = 0 & \text{in } \Omega_T, \\ u = 0, \quad H \cdot n = 0, \quad \operatorname{rot} H \times n = 0 & \text{on } \partial\Omega_T, \end{cases}$$

where we used the abbreviations $S_p^1(H) = (H_0 \cdot \nabla)H + (H \cdot \nabla)H_0$, $F(H_0) = F + (H_0 \cdot \nabla)H_0$, and added $\frac{1}{2}\nabla|H + H_0|^2$ to the pressure. We are going to show existence of solutions to these equations by a fixed point argument, based on the results of Theorem 4.3.1. As a first step we provide estimates of the nonlinearities in the following lemma.

Lemma 4.4.1. *Let* $1 < p,q < \infty$ *such that* $\frac{2}{q} + \frac{3}{p} \leq 3$ *and* $u, v \in W_{q,p}^{1,2}(\Omega_T)$. *Then it holds*

$$\|(u \cdot \nabla)v\|_{L_{q,p}(\Omega_T)} \lesssim \|u\|_{W_{q,p}^{1,2}(\Omega_T)}\|v\|_{W_{q,p}^{1,2}(\Omega_T)}.$$

Proof: The idea is to apply Hölder's inequality with $\frac{1}{q_0} + \frac{1}{q_1} = \frac{1}{q}$ and $\frac{1}{p_0} + \frac{1}{p_1} = \frac{1}{p}$ where q_i, p_i with $i \in \{1,2\}$ are chosen appropriately such that Corollary 3.1.35 is applicable and yields

$$\|(u \cdot \nabla)u\|_{L_{q,p}(\Omega_T)} \leq \|u\|_{L_{q_0,p_0}(\Omega_T)}\|\nabla v\|_{L_{q_1,p_1}(\Omega_T)} \leq \|u\|_{W_{q,p}^{1,2}(\Omega_T)}\|v\|_{W_{q,p}^{1,2}(\Omega_T)}.$$

Since the condition $\frac{2}{q} + \frac{3}{p} \leq 3$ is equivalent to $q \geq \frac{2p}{3p-3}$ for $p > 1$ we consider the following three cases.

- For $p \geq 3$ and $q > 1$ we consider $\alpha = \beta = \frac{1}{q} < 1$ to obtain $q_0 = 2q$, $q_1 = 2q$, $p_0 = \infty$ and $p_1 = p$, which covers all possible combinations.

- For $p < 3$ and $q > \frac{2p}{3p-3}$ we consider $\alpha = \frac{3p-3}{p} < 2$ and conclude $\alpha q > 2$. Hence, choosing $\beta = 0$ as well as $q_0 = \infty$, $p_0 = \frac{3p}{3-(2-\alpha)p}$, $q_1 = q$ and $p_1 = \frac{3p}{3-p}$ we deduce $\frac{1}{p_0} + \frac{1}{p_1} = \frac{1}{p}$.

- For $\frac{2}{q} + \frac{3}{p} = 3$ we consider $\alpha = \beta = \frac{1}{q}$ and obtain $\frac{1}{q_0} + \frac{1}{q_1} = \frac{1}{q}$, similar to the first case, and

$$\frac{1}{p_0} + \frac{1}{p_1} = \frac{6 - 3p + \frac{2p}{q}}{3p} = \frac{1}{p}.$$

Hence the estimate holds for all stated p, q. $\qquad\square$

Remark 4.4.2. Note that the condition $\frac{2}{q} + \frac{3}{p} \leq 3$ is justified by the embeddings of Corollary 3.1.35, because it occurs naturally from assuming the worst embeddings for all p_i, q_i. A straightforward calculation shows that in fact with the stated embeddings no other combinations are possible. For example, assuming $\frac{2}{q} + \frac{3}{p} > 3$ and $\alpha \geq \frac{2}{q}$ yields $(2 - \alpha)p < 3$ and $(1 - \beta)p \geq 3$, which already is a contradiction since it implies $\frac{2}{q} + \frac{3}{p} \leq 3$. The other cases follow similarly.

With this preparation we can state and prove the main existence theorem.

Theorem 4.4.3. *Let* $1 < p,q < \infty$ *such that* $\frac{2}{q} + \frac{3}{p} \leq 3$. *To every* $\kappa > 0$ *and* $\|H_0\|_{W_\infty^1(\Omega)} < \kappa$ *there exists an* $\varepsilon > 0$ *such that for all* $F \in L_{q,p}(\Omega_T)$ *satisfying*

$$\|F + (H_0 \cdot \nabla)H_0\|_{L_{q,p}(\Omega_T)} \leq \varepsilon$$

there exists a unique solution $(u, H, \mathfrak{p}) \in W_{q,p}^{1,2}(\Omega_T)^2 \times \widehat{W}_{q,p}^{0,1}(\Omega_T)$ *to (4.36).*

Proof: We are going to show existence of a solution by Banach's fixed point theorem. To apply this theorem we define the operator

$$\mathfrak{L} : Y_{q,p}^{1,2}(\Omega_T) \times X_{q,p}^{1,2}(\Omega_T) \times \widehat{W}_{q,p}^{0,1}(\Omega_T) \rightarrow Y_{q,p}^{1,2}(\Omega_T) \times X_{q,p}^{1,2}(\Omega_T) \times \widehat{W}_{q,p}^{0,1}(\Omega_T)$$

$$\mathfrak{L}(u, H, \mathfrak{q}) := Z_{H_0}^{-1}\begin{pmatrix} F(H_0) + (H \cdot \nabla)H - (u \cdot \nabla)u \\ \nabla \times [u \times H] \end{pmatrix},$$

with $Z_{H_0}^{-1}$ as the solution operator given by Theorem 4.3.1. By the same theorem we know that $Z_{H_0}^{-1}$ is a continuous operator with norm bounded by a constant $c(\kappa)$, independent of H_0 as long as $\|H_0\|_{W_\infty^1(\Omega)} < \kappa$. Hence it holds

$$\|\mathfrak{L}(u, H, \mathfrak{q})\|_{W_{q,p}^{1,2}(\Omega_\mathrm{T})} \leq c(\kappa)\Big(\|F(H_0)\|_{L_{q,p}(\Omega_\mathrm{T})} + \|(H \cdot \nabla)H\|_{L_{q,p}(\Omega_\mathrm{T})} + \|(u \cdot \nabla)u\|_{L_{q,p}(\Omega_\mathrm{T})}$$
$$+ \|\nabla \times [u \times H]\|_{L_{q,p}(\Omega_\mathrm{T})}\Big)$$

Since $\nabla \times [u \times H] = (H \cdot \nabla)u - (u \cdot \nabla)H$, we can apply Lemma 4.4.1 to obtain

$$(4.37) \quad \|\mathfrak{L}(u, H, \mathfrak{q})\|_{W_{q,p}^{1,2}(\Omega_\mathrm{T})} \leq c(\kappa)\Big(\|F + (H_0 \cdot \nabla)H_0\|_{L_{q,p}(\Omega_\mathrm{T})} + \Big[\|H\|_{W_{q,p}^{1,2}(\Omega_\mathrm{T})} + \|u\|_{W_{q,p}^{1,2}(\Omega_\mathrm{T})}\Big]^2\Big).$$

We want to show that $\mathfrak{L}$ is a contracting self-mapping on balls with sufficiently small radius $\delta > 0$. Choosing $(u, H) \in \big[Y_{q,p}^{1,2}(\Omega_\mathrm{T}) \times X_{q,p}^{1,2}(\Omega_\mathrm{T})\big] \cap \overline{B_\delta}$, $\delta \leq \frac{1}{2c(\kappa)}$ and $\varepsilon \leq \frac{\delta}{2c(\kappa)}$, we conclude from (4.37)

$$\|\mathfrak{L}(u, H, \mathfrak{q})\|_{W_{q,p}^{1,2}(\Omega_\mathrm{T})} \leq c(\kappa)\Big(\frac{\delta}{2c(\kappa)} + \delta^2\Big) \leq \delta,$$

hence $\mathfrak{L}$ is a self-mapping on $\Big(\big[Y_{q,p}^{1,2}(\Omega_\mathrm{T}) \times X_{q,p}^{1,2}(\Omega_\mathrm{T})\big] \cap \overline{B_\delta}\Big) \times \widehat{W}_{q,p}^{0,1}(\Omega_\mathrm{T})$. Furthermore we obtain

$$\|\mathfrak{L}(u_1, H_1, \mathfrak{q}_1) - \mathfrak{L}(u_2, H_2, \mathfrak{q}_2)\|_{W_{q,p}^{1,2}(\Omega_\mathrm{T})} \leq c(\kappa)\|\nabla H_1 \cdot H_1 - \nabla H_2 \cdot H_2\|_{L_{q,p}(\Omega_\mathrm{T})}$$
$$+ c(\kappa)\big[\|\nabla u_1 \cdot u_1 - \nabla u_2 \cdot u_2\|_{L_{q,p}(\Omega_\mathrm{T})} + \|\nabla H_1 \cdot u_1 - \nabla H_2 \cdot u_2\|_{L_{q,p}(\Omega_\mathrm{T})}\big]$$
$$+ c(\kappa)\|\nabla u_1 \cdot H_1 - \nabla u_2 \cdot H_2\|_{L_{q,p}(\Omega_\mathrm{T})}.$$

Since all terms above have the same structure we are going to show the estimate for one, because the rest easily follows by the same idea. We use the estimates of Lemma 4.4.1 to obtain

$$\|\nabla H_1 \cdot u_1 - \nabla H_2 \cdot u_2\|_{L_{q,p}(\Omega_\mathrm{T})} \leq \|\nabla H_1 \cdot u_1 - \nabla H_1 \cdot u_2\|_{L_{q,p}(\Omega_\mathrm{T})} + \|\nabla H_1 \cdot u_2 - \nabla H_2 \cdot u_2\|_{L_{q,p}(\Omega_\mathrm{T})}$$
$$\leq \|H_1\|_{W_{q,p}^{1,2}(\Omega_\mathrm{T})}\|u_1 - u_2\|_{W_{q,p}^{1,2}(\Omega_\mathrm{T})} + \|u_2\|_{W_{q,p}^{1,2}(\Omega_\mathrm{T})}\|H_1 - H_2\|_{W_{q,p}^{1,2}(\Omega_\mathrm{T})}$$
$$\leq \delta\big(\|u_1 - u_2\|_{W_{q,p}^{1,2}(\Omega_\mathrm{T})} + \|H_1 - H_2\|_{W_{q,p}^{1,2}(\Omega_\mathrm{T})}\big).$$

Combining all previous estimates yields

$$\|\mathfrak{L}(u_1, H_1, \mathfrak{q}_1) - \mathfrak{L}(u_2, H_2, \mathfrak{q}_2)\|_{W_{q,p}^{1,2}(\Omega_\mathrm{T})} \leq 4\delta c(\kappa)\Big(\|u_1 - u_2\|_{W_{q,p}^{1,2}(\Omega_\mathrm{T})} + \|H_1 - H_2\|_{W_{q,p}^{1,2}(\Omega_\mathrm{T})}\Big).$$

Hence by restricting δ further to $\delta \leq \frac{1}{5c(\kappa)}$ we derive

$$\|\mathfrak{L}(u_1, H_1, \mathfrak{q}_1) - \mathfrak{L}(u_2, H_2, \mathfrak{q}_2)\|_{W_{q,p}^{1,2}(\Omega_\mathrm{T})} \leq \frac{4}{5}\|(u_1, H_1, \mathfrak{q}_1) - (u_2, H_2, \mathfrak{q}_2)\|_{W_{q,p}^{1,2}(\Omega_\mathrm{T})},$$

and therefore there exists a unique fixed point $(u, H, \mathfrak{q}) \in Y_{q,p}^{1,2}(\Omega_\mathrm{T}) \times X_{q,p}^{1,2}(\Omega_\mathrm{T}) \times \widehat{W}_{q,p}^{0,1}(\Omega_\mathrm{T})$ and thus a solution to (4.36). $\qquad\square$

We collect the results of this chapter in the following corollary, which will state the main existence result of time-periodic solutions to the equations of magnetohydrodynamics (MHD).

Corollary 4.4.4. *Let $1 < p, q < \infty$ such that $\frac{2}{q} + \frac{3}{p} \leq 3$ and $\delta > 0$. For every $B_1 \in \mathrm{W}_r^{2-\frac{1}{r}}(\partial\Omega)$ with $r > 3$ and $r \geq p$ such that*

$$\text{(4.38)} \qquad \int\limits_{\partial\Omega} B_1(x) \, \mathrm{d}\sigma = 0 \quad and \quad \|B_1\|_{\mathrm{W}_r^{2-\frac{1}{r}}(\partial\Omega)} < \delta,$$

there exists an $\varepsilon > 0$ such that for all $F \in \mathrm{L}_{q,p}(\Omega_\mathbb{T})$ satisfying

$$\text{(4.39)} \qquad \|F + (H_0 \cdot \nabla)H_0\|_{\mathrm{L}_{q,p}(\Omega_\mathbb{T})} \leq \varepsilon$$

there exists a solution $(u, H, \mathfrak{p}) \in \mathrm{W}_{q,p}^{1,2}(\Omega_\mathbb{T})^2 \times \mathrm{W}_{q,p}^{0,1}(\Omega_\mathbb{T})$ to (MHD). Here H_0 is the extension of B_1 constructed by (4.1).

Proof: For every B_1 satisfying (4.38) there exists an H_0 such that $H_0 \cdot n = B_1$ on $\partial\Omega$ and $\|H_0\|_{\mathrm{W}_\infty^1(\Omega)} \lesssim \|B_1\|_{\mathrm{W}_r^{2-\frac{1}{r}}(\partial\Omega)}$ by solving (4.1) and using (4.2). Hence we transform (MHD) into (4.36) via $H = H_1 + H_0$. Therefore, we obtain by Theorem 4.4.3 a solution $(u_1, H_1, \mathfrak{q}) \in \mathrm{W}_{q,p}^{1,2}(\Omega_\mathbb{T})^2 \times \widehat{\mathrm{W}}_{q,p}^{0,1}(\Omega_\mathbb{T})$ to (4.36) for all $F \in \mathrm{L}_{q,p}(\Omega_\mathbb{T})$ satisfying (4.39) with suitable $\varepsilon > 0$ depending on δ. We set $(u, H, \mathfrak{p}) = (u_1, H_1 + H_0, \mathfrak{q} + \frac{1}{2}|H_1 + H_0|^2)$ and obtain a solution to (MHD). Since $|H_1 + H_0|^2$ is an element of $\mathrm{W}_{q,p}^{0,1}(\Omega_\mathbb{T})$ by Lemma 4.4.1 and $H_0 \in \mathrm{W}_\infty^1(\Omega)$, by choosing the $\mathfrak{q}$ with mean value zero we employ Poincaré's inequality to obtain $\mathfrak{q} \in \mathrm{W}_{q,p}^{0,1}(\Omega_\mathbb{T})$ and hence $\mathfrak{p} \in \mathrm{W}_{q,p}^{0,1}(\Omega_\mathbb{T})$. Because $H_0 \in \mathrm{W}_r^2(\Omega)$ with $r \geq p$ the regularity of H is obvious and hence the results follows. $\square$

Remark 4.4.5. The smallness assumptions in Corollary 4.4.4 and Theorem 4.4.3 may seem a bit odd, since they depend on the choice of extension of the boundary data B_1. But as stated in the introduction, the magnetic field on the boundary is the intrinsic field of the medium containing the boundary. So if for example the data B_1 is given by $B_0 \cdot n$ for any constant $B_0 \in \mathbb{R}$, then (4.38) is satisfied and (4.39) reduces to smallness of F since all other terms vanish because H_0 is constant. Hence, Corollary 4.4.4 implies that for every constant magnetic field H_0 in the background there exists a strong time-periodic solution to (MHD). In the case where H_0 is not constant we are still able to show existence of solutions if either H_0 is small or the change in the magnetic field compared to the magnetic field is small, *i.e.*, the expression $\nabla H_0 \cdot H_0$ is small, see (4.39).

Remark 4.4.6. The regularity assumption of B_1 is natural in the case of $p > 3$, since we know that the trace $H \cdot n$ is an element of $\mathrm{W}_p^{2-\frac{1}{p}}(\partial\Omega)$ by standard theory. So only in the case of $p \leq 3$ it is necessary to demand additional regularity.

CHAPTER 5

Appendix

5.1 Auxiliary Results

In this section we will provide a variety of useful results, ranging from estimates over identities to properties of some function spaces. We start with an identity needed for an application of Theorem 1.3.1.

Lemma 5.1.1. *Let $f_j \in \mathrm{C}^1(\mathbb{R})$ be positive functions for $j = 1, 2, \ldots, n$, $g \in \mathrm{C}^n(\mathbb{R}^n)$ and set $h(x) = 1 + \sum_{j=1}^{n} f_j(x_j)$ for all $x \in \mathbb{R}^n$. Then for every $\alpha \in \{0, 1\}^n$ the identity*

$$D^\alpha \frac{g(x)}{h(x)} = \sum_{k=0}^{|\alpha|} \sum_{\substack{\gamma \in \mathbb{N}_0^n, \\ \gamma \leq \alpha, \\ |\gamma| = k}} (-1)^k k! \frac{(D^{\alpha-\gamma} g(x)) \prod_{j=1}^{n} \left(f_j'(x_j)\right)^{\gamma_j}}{h(x)^{1+k}}$$

holds. The second sum is over all multi-indices $\gamma \in \mathbb{N}_0^n$ with the stated properties.

Proof: We prove this by induction over $|\alpha|$. For $|\alpha| = 0$ the identity holds true since the sum has only one element. We assume that the identity holds for some $m \in \mathbb{N}_0$ and take $|\alpha| = m + 1$. Therefore, we find $\alpha' \in \mathbb{N}_0^n$ with $|\alpha'| = m$ and $\alpha = \alpha' + e_l$ for some $l \in \{1, 2, \ldots, n\}$. Hence we obtain

$$D^\alpha \frac{g(x)}{h(x)} = D^{\alpha'+e_l} \frac{g(x)}{h(x)} = D^{e_l} \left[\sum_{k=0}^{|\alpha'|} \sum_{\substack{\gamma \in \mathbb{N}_0^n, \\ \gamma \leq \alpha', \\ |\gamma|=k}} (-1)^k k! \frac{\left(D^{\alpha'-\gamma} g(x)\right) \prod_{j=1}^n \left(f_j'(x_j)\right)^{\gamma_j}}{h(x)^{1+k}} \right]$$

$$= \sum_{k=0}^{|\alpha'|} \sum_{\substack{\gamma \in \mathbb{N}_0^n, \\ \gamma \leq \alpha', \\ |\gamma|=k}} (-1)^k k! \frac{\left(D^{\alpha-\gamma} g(x)\right) \prod_{j=1}^n \left(f_j'(x_j)\right)^{\gamma_j}}{h(x)^{1+k}}$$

$$+ \sum_{k=0}^{|\alpha'|} \sum_{\substack{\gamma \in \mathbb{N}_0^n, \\ \gamma \leq \alpha', \\ |\gamma|=k}} (-1)^{k+1} (k+1)! \frac{\left(D^{\alpha-\gamma-e_l} g(x)\right) \prod_{j=1}^n \left(f_j'(x_j)\right)^{\gamma_j} f_l'(x_l)}{h(x)^{2+k}}$$

$$= \sum_{k=0}^{|\alpha|} \sum_{\substack{\gamma \in \mathbb{N}_0^n, \\ \gamma \leq \alpha, \\ |\gamma|=k}} (-1)^k k! \frac{\left(D^{\alpha-\gamma} g(x)\right) \prod_{j=1}^n \left(f_j'(x_j)\right)^{\gamma_j}}{h(x)^{2+k}}$$

Since the functions f_j only depend on one variable and $\alpha \in \{0,1\}^n$ no second derivatives of f_j occur. The last identity follows by comparing the elements for fixed k. In the first sum the function g has at least a derivative with respect to x_l, in the second sum g will never be differentiated with respect to x_l. $\qquad\square$

The following result is well-known for homogeneous functions of any degree and we extend it to anisotropic scaling.

Lemma 5.1.2. Let $\vec{a} \in (0,\infty)^n$ and $f \in \mathbb{C}^\infty(\mathbb{R}^n \setminus \{0\})$ with anisotropic scaling, i.e., $f(t^{\vec{a}} x) = t^s f(x)$ for arbitrary $s \in \mathbb{R}$ and $x \in \mathbb{R}^n \setminus \{0\}$. Then for any $\alpha \in \mathbb{N}_0^n$ it holds

$$(5.1) \qquad\qquad (D^\alpha f)(t^{\vec{a}} x) = t^{s - \vec{a} \cdot \alpha} D^\alpha f(x).$$

Proof: We are going to prove this by induction over $|\alpha|$. Let $\alpha \in \mathbb{N}_0^n$ be arbitrary. We find $j \in \{1, 2, \ldots, n\}$ and $\gamma \in \mathbb{N}_0^n$ so that $\gamma + e_j = \alpha$ and $|\gamma| = |\alpha| - 1$. For the base case γ is just zero. Therefore, it holds

$$(D^\alpha f)(t^{\vec{a}} x) = (D^{e_j} D^\gamma f)(t^{\vec{a}} x) = \lim_{h \to 0} \frac{(D^\gamma f)(t^{a_1} x_1, \ldots, t^{a_j} x_j + h, \ldots, t^{a_n} x_n) - (D^\gamma f)(t^{\vec{a}} x)}{h}$$

$$= \lim_{h \to 0} \frac{(D^\gamma f)(t^{a_1} x_1, \ldots, t^{a_j}(x_j + \frac{h}{t^{a_j}}), \ldots, t^{a_n} x_n) - (D^\gamma f)(t^{\vec{a}} x)}{t^{a_j} \frac{h}{t^{a_j}}}$$

$$= \lim_{h \to 0} \frac{t^{s - \gamma \cdot \vec{a}} \left[(D^\gamma f)(x_1, \ldots, x_j + \frac{h}{t^{a_j}}, \ldots, x_n) - (D^\gamma f)(x) \right]}{t^{a_j} \frac{h}{t^{a_j}}}$$

$$= t^{s - \gamma \cdot \vec{a} - e_j \cdot \vec{a}} \lim_{h \to 0} \frac{(D^\gamma f)(x_1, \ldots, x_j + h, \ldots, x_n) - (D^\gamma f)(x)}{h}$$

$$= t^{s - \vec{a} \cdot \alpha} D^\alpha f(x),$$

which proves the assertion. $\qquad\square$

As stated in the beginning of Section 2.1 the following inequality of the Dirichlet kernel D_K is fundamental.

Lemma 5.1.3. *Let $K \in \mathbb{N}$, $1 < p < \infty$ and $D_K : \mathbb{T} \to \mathbb{C}$ with $D_K(t) = \frac{\sin([\frac{2\pi}{\mathcal{T}}K + \frac{\pi}{\mathcal{T}}]t)}{\sin(t\frac{\pi}{\mathcal{T}})}$. Then there exists a constant $c_p > 0$ such that*

$$\|D_K\|_{L_p(\mathbb{T})} \leq c_p K^{1-\frac{1}{p}}$$

holds.

Proof: To simplify the notation we set $R = \frac{2\pi}{\mathcal{T}}K + \frac{\pi}{\mathcal{T}}$. On $[-\frac{\mathcal{T}}{2}, \frac{\mathcal{T}}{2}]$ we have the estimate $|\frac{t\pi}{2\mathcal{T}}| \leq |\sin(t\frac{\pi}{\mathcal{T}})|$ and since D_K is $\mathcal{T}$-periodic we conclude

$$\left(\frac{2\pi}{\mathcal{T}}K + \frac{\pi}{\mathcal{T}}\right)^{1-p} \|D_K\|_{L_p(\mathbb{T})}^p = R^{1-p} \int_{\mathbb{T}} \left|\frac{\sin(Rt)}{\sin(t\frac{\pi}{\mathcal{T}})}\right|^p dt \leq R^{1-p} \int_{-\frac{\mathcal{T}}{2}}^{\frac{\mathcal{T}}{2}} \left|\frac{\sin(Rt)}{\frac{t\pi}{2\mathcal{T}}}\right|^p dt$$

$$= R^{1-p} \int_{-R\frac{\mathcal{T}}{2}}^{R\frac{\mathcal{T}}{2}} \frac{1}{R} \left|\frac{\sin(s)}{\frac{s\pi}{R2\mathcal{T}}}\right|^p ds$$

$$\leq \left(\frac{2\mathcal{T}}{\pi}\right)^p \int_{\mathbb{R}} \left|\frac{\sin(s)}{s}\right|^p ds =: c_p.$$

The last integral converges because $p > 1$ and the integrand is continuous on $\mathbb{R}$. $\qquad \square$

We collect some estimates regarding sequences in the following results. The first extends the result of Brezis and Mironescu [18, Lemma 3.7] from $R = 2$ to arbitrary $R > 1$.

Lemma 5.1.4. *Let $-\infty < s_1 < s_2 < \infty$, $R > 1$, $0 < q \leq \infty$ and $0 < \theta < 1$. Then there exists a constant $c(R) > 0$ such that for every sequence $\{a_j\}_{j \in \mathbb{N}_0}$ we have*

$$\|\{R^{sj}a_j\}_{j \in \mathbb{N}_0}\|_{l_q(\mathbb{N}_0)} \leq c \|\{R^{s_1 j}a_j\}_{j \in \mathbb{N}_0}\|_{l_\infty(\mathbb{N}_0)}^\theta \|\{R^{s_2 j}a_j\}_{j \in \mathbb{N}_0}\|_{l_\infty(\mathbb{N}_0)}^{1-\theta},$$

with $s = \theta s_1 + (1-\theta)s_2$.

Proof: The case of $q = \infty$ is obvious, so we consider $q < \infty$ and furthermore we can assume that the right hand side is finite, because otherwise the inequality is obviously true. We define $C_1 = \|R^{s_1 j}a_j\|_{l_\infty}$ and $C_2 = \|R^{s_2 j}a_j\|_{l_\infty}$. Because $s_1 < s_2$ we have $C_1 \leq C_2$ and thus find a j_0 such that

$$(5.2) \qquad\qquad R^{j_0(s_2-s_1)} \leq \frac{C_2}{C_1} < R^{(j_0+1)(s_2-s_1)}.$$

By definition $R^{s_i j}|a_j| \leq C_i$ and hence $|a_j| \leq C_i R^{-s_i j}$ for all $j \in \mathbb{N}_0$ and $i = 1, 2$. This implies

$$\|\{R^{sj}a_j\}_{j \in \mathbb{N}_0}\|_{l_q}^q = \sum_{j \leq j_0} R^{sjq}|a_j|^q + \sum_{j > j_0} R^{sjq}|a_j|^q \leq \sum_{j \leq j_0} R^{(s-s_1)jq}C_1^q + \sum_{j > j_0} R^{(s-s_2)jq}C_2^q$$

$$\leq \sum_{j \leq j_0} R^{(s-s_1)jq}C_1^q + R^{(s_2-s_1)q} \sum_{j > j_0} R^{(s-s_2)jq}C_1^q R^{j_0(s_2-s_1)q}$$

$$\leq R^{(s_2-s_1)q}C_1^q \left[\sum_{j \leq j_0} R^{(s-s_1)jq} + \sum_{j > j_0} R^{(s-s_2)jq} R^{j_0(s_2-s_1)q}\right]$$

$$= C_1^q R^{(s_2-s_1)q[j_0(1-\theta)+1]} \left[\sum_{j \leq j_0} \frac{R^{(s-s_1)jq}}{R^{(s_2-s_1)j_0(1-\theta)q}} + \sum_{j > j_0} R^{(s-s_2)jq+j_0(s_2-s_1)\theta q}\right].$$

The definition of s implies $s - s_1 = (s_2 - s_1)(1 - \theta)$ and $s - s_2 = -\theta(s_2 - s_1)$. Applying these identities together with (5.2) yields

$$\|\{R^{sj}a_j\}_{j\in\mathbb{N}_0}\|_{l_q}^q \leq C_1^q R^{(s_2-s_1)q[j_0(1-\theta)+1]} \left[\sum_{j\leq j_0} R^{(s_2-s_1)(j-j_0)(1-\theta)q} + \sum_{j>j_0} R^{q\theta(s_2-s_1)(j_0-j)} \right]$$

$$\lesssim C_1^q R^{(s_2-s_1)j_0(1-\theta)q} \leq C_1^q \frac{C_2^{q(1-\theta)}}{C_1^{q(1-\theta)}} = C_1^{q\theta} C_2^{q(1-\theta)},$$

which proves the assertion. $\qquad\square$

The following lemma is stated in Johnsen and Sickel [56, Lemma 4.2] and in Johnsen [52, Lemma 2.5] but without a proof, which we will give here. It extends the ideas of Yamazaki [85, Lemma 3.8.]

Lemma 5.1.5. *Let $s \in \mathbb{R}$ with $s > 0$ and $q, r \in [1, \infty]$. Then there exists a constant $c = c(s, q, r)$ such that for every sequence $(b_j)_{j\in\mathbb{N}_0} \subset \mathbb{C}$ it holds*

$$\left\| \left\{ 2^{sj} \left(\sum_{k=j}^{\infty} |b_k|^r \right)^{\frac{1}{r}} \right\}_{j\in\mathbb{N}_0} \right\|_{\ell_q} \leq c \left\| \{2^{sj}b_j\}_{j\in\mathbb{N}_0} \right\|_{\ell_q},$$

$$\left\| \left\{ 2^{-sj} \left(\sum_{k=0}^{j} |b_k|^r \right)^{\frac{1}{r}} \right\}_{j\in\mathbb{N}_0} \right\|_{\ell_q} \leq c \left\| \{2^{-sj}b_j\}_{j\in\mathbb{N}_0} \right\|_{\ell_q}.$$

For $r = \infty$ the sum has to be exchanged with the supremum.

Proof: We first consider $1 \leq q \leq r < \infty$. By the monotonicity of the ℓ_r-spaces and Fubini's theorem we have

$$\left\| \left\{ 2^{sj} \left(\sum_{k=j}^{\infty} |b_k|^r \right)^{\frac{1}{r}} \right\}_{j\in\mathbb{N}} \right\|_{\ell_q}^q \leq \sum_{j=0}^{\infty} 2^{sjq} \sum_{k=j}^{\infty} |b_k|^q = \sum_{k=0}^{\infty} 2^{skq}|b_k|^q \sum_{j=0}^{k} 2^{s(j-k)q}$$

$$\lesssim \|\{2^{sj}b_j\}_{j\in\mathbb{N}_0}\|_{\ell_q}^q.$$

For $r = \infty$ or $r = q = \infty$ the proof holds true, if one exchanges the sums for supremums. For $1 \leq r < q < \infty$ we apply Hölder's inequality with $1 < \frac{q}{r} < \infty$ to the ℓ_1-norm of $|b_k|^r$ to obtain

$$\left\| \left\{ 2^{sj} \left(\sum_{k=j}^{\infty} |b_k|^r \right)^{\frac{1}{r}} \right\}_{j\in\mathbb{N}_0} \right\|_{\ell_q}^q \leq \sum_{j=0}^{\infty} \sum_{k=j}^{\infty} 2^{sjq}|b_k|^q 2^{s(k-j)\frac{q}{2}} \left(\sum_{k=j}^{\infty} 2^{-s(k-j)\frac{qr}{(q-r)2}} \right)^{\frac{q-r}{r}}$$

$$= \sum_{k=0}^{\infty} 2^{skq}|b_k|^q \sum_{j=0}^{k} 2^{s(j-k)\frac{q}{2}} \left(\sum_{k=0}^{\infty} 2^{-sk\frac{qr}{(q-r)2}} \right)^{\frac{q-r}{r}}$$

$$\lesssim \|\{2^{sj}b_j\}_{j\in\mathbb{N}_0}\|_{\ell_q}^q.$$

For $q = \infty$ we derive

$$\left\| \left\{ 2^{sj} \left(\sum_{k=j}^{\infty} |b_k|^r \right)^{\frac{1}{r}} \right\}_{j\in\mathbb{N}} \right\|_{\ell_\infty} \leq \sup_{j\in\mathbb{N}_0} 2^{sj} \left(\sup_{k\in\mathbb{N}_0} 2^{ksr}|b_k|^r \sum_{k=j}^{\infty} 2^{-(k-j)sr}2^{-jsr} \right)^{\frac{1}{r}}$$

$$= \sup_{k\in\mathbb{N}_0} 2^{ks}|b_k| \sup_{j\in\mathbb{N}_0} \left(\sum_{k=j}^{\infty} 2^{-(k-j)sr} \right)^{\frac{1}{r}} = c \|\{2^{ks}b_k\}_{k\in\mathbb{N}_0}\|_{\ell_\infty}.$$

This proves the first assertion for all $p, r \in [1, \infty]$.

To prove the second estimate we start with $1 \le q \le r < \infty$ and use the monotonicity of the ℓ_r-spaces and Fubini's theorem to obtain

$$\left\| \left\{ 2^{-sj} \left(\sum_{k=0}^{j} |b_k|^r \right)^{\frac{1}{r}} \right\}_{j \in \mathbb{N}_0} \right\|_{\ell_q}^q \le \sum_{j=0}^{\infty} 2^{-sjq} \sum_{k=0}^{j} |b_k|^q = \sum_{k=0}^{\infty} 2^{-skq} |b_k|^q \sum_{j=k}^{\infty} 2^{-s(j-k)q}$$
$$\lesssim \left\| \{ 2^{-sj} b_j \}_{j \in \mathbb{N}_0} \right\|_{\ell_q}^q.$$

For $1 \le r < q < \infty$ we repeat the previous ideas and apply Hölder's inequality with $1 < \frac{q}{r} < \infty$ to the ℓ_1 of $|b_k|^r$ and conclude

$$\left\| \left\{ 2^{-sj} \left(\sum_{k=0}^{j} |b_k|^r \right)^{\frac{1}{r}} \right\}_{j \in \mathbb{N}_0} \right\|_{\ell_q}^q \le \sum_{j=0}^{\infty} \sum_{k=0}^{j} 2^{-sjq} |b_k|^q 2^{s(j-k)\frac{q}{2}} \left(\sum_{k=0}^{j} 2^{-s(j-k)\frac{qr}{(q-r)2}} \right)^{\frac{q-r}{r}}$$
$$\le \sum_{k=0}^{\infty} 2^{-skq} |b_k|^q \sum_{j=k}^{\infty} 2^{s(k-j)\frac{q}{2}} \left(\sum_{k=0}^{\infty} 2^{-sk\frac{qr}{(q-r)2}} \right)^{\frac{q-r}{r}}$$
$$= \left\| \{ 2^{-sj} b_j \}_{j \in \mathbb{N}_0} \right\|_{\ell_q}^q.$$

Repeating the previous arguments includes $q = \infty$. Exchanging the sums for supremums includes $r = \infty$ and $q = r = \infty$ and hence the result. $\qquad\qquad\qquad\square$

As stated in the beginning of Section 3.1 the following result is justification for calling $\widehat{B}_{\vec{p},r}^{s,\vec{a}}$ homogeneous Besov spaces.

Lemma 5.1.6. *Let $1 \le \vec{p}, r \le \infty$, let $\vec{a} \in (0, \infty)^n$ and $\frac{\vec{a}}{\vec{p}} = \sum_{j=1}^{n} \frac{a_j}{p_j}$. Then it holds*

$$\left\| u(2^{k\vec{a}} \cdot) \right\|_{\widehat{B}_{\vec{p},r}^{s,\vec{a}}(\mathbb{R}^n)} = 2^{kr(s - \frac{\vec{a}}{\vec{p}})} \| u \|_{\widehat{B}_{\vec{p},r}^{s,\vec{a}}(\mathbb{R}^n)}$$

for all $k \in \mathbb{Z}$ if either of the norms is finite for $u \in \mathscr{S}'(\mathbb{R}^n)$.

Proof: As a first step we calculate some identities. For $f \in \mathscr{S}(\mathbb{R}^n)$, $g \in \mathscr{S}(\mathbb{R})$ and $a \in \mathbb{R}$ it holds

$$\mathscr{F}_{\mathbb{R}^n}[f(2^{j\vec{a}} \cdot)](\xi) = \int_{\mathbb{R}^n} f(2^{j\vec{a}} x) e^{ix \cdot \xi} \, dx = 2^{-j|\vec{a}|} \int_{\mathbb{R}^n} f(y) e^{iy \cdot [2^{-j\vec{a}} \xi]} \, dy = 2^{-j|\vec{a}|} \mathscr{F}_{\mathbb{R}^n}[f](2^{-j\vec{a}} \xi),$$
$$\| g(2^{ka} \cdot) \|_{L_p(\mathbb{R})} = 2^{\frac{-ka}{p}} \| g \|_{L_p(\mathbb{R})},$$

with obvious modifications for $p = \infty$. The first identity extends by duality to $u \in \mathscr{S}'(\mathbb{R}^n)$ and applying the second one inductively yields

$$(5.3) \qquad \mathscr{F}_{\mathbb{R}^n}[u(2^{j\vec{a}} \cdot)] = 2^{-j|\vec{a}|} \mathscr{F}_{\mathbb{R}^n}[u](2^{-j\vec{a}} \cdot), \quad \mathscr{F}_{\mathbb{R}^n}^{-1}[u(2^{j\vec{a}} \cdot)] = 2^{-j|\vec{a}|} \mathscr{F}_{\mathbb{R}^n}^{-1}[u](2^{-j\vec{a}} \cdot),$$

$$(5.4) \qquad \| f(2^{k\vec{a}} \cdot) \|_{L_{\vec{p}}(\mathbb{R}^n)} = 2^{-k\frac{\vec{a}}{\vec{p}}} \| f \|_{L_{\vec{p}}(\mathbb{R}^n)}.$$

Therefore, by the properties of ϕ_k, see (2.27), we have

$$\mathscr{F}^{-1}[\phi_j(\xi) \mathscr{F}[u(2^{k\vec{a}} \xi)]](x) = 2^{-k|\vec{a}|} \mathscr{F}^{-1}[\phi_j(\xi)(\mathscr{F}u)(2^{-k\vec{a}} \xi)](x)$$
$$= 2^{-k|\vec{a}|} \mathscr{F}^{-1}[\phi_{j-k}(2^{-k\vec{a}} \xi)(\mathscr{F}u)(2^{-k\vec{a}} \xi)](x)$$
$$= \mathscr{F}^{-1}[\phi_{j-k}(\xi) \mathscr{F}u(\xi)](2^{k\vec{a}} x).$$

We combine the previous results, so that the identity

$$\sum_{j\in\mathbb{Z}} 2^{jsr} \left\| \mathscr{F}^{-1}\left[\phi_j \mathscr{F}[u(2^{k\vec{a}}\cdot)]\right](\cdot) \right\|_{\vec{p}}^r = \sum_{j\in\mathbb{Z}} 2^{jsr} \left\| \mathscr{F}^{-1}\left[\phi_{j-k}\mathscr{F}u\right](2^{k\vec{a}}\cdot) \right\|_{\vec{p}}^r$$

$$= \sum_{j\in\mathbb{Z}} 2^{(j-k)sr+ksr-kr\frac{\vec{a}}{\vec{p}}} \left\| \mathscr{F}^{-1}\left[\phi_{j-k}\mathscr{F}u\right](\cdot) \right\|_{\vec{p}}^r$$

$$= 2^{kr(s-\frac{\vec{a}}{\vec{p}})} \sum_{j\in\mathbb{Z}} 2^{jsr} \left\| \mathscr{F}^{-1}\left[\phi_j \mathscr{F}u\right](\cdot) \right\|_{\vec{p}}^r$$

holds. □

In the isotropic case it is known, that L_p-regularity of a function in a homogeneous Besov $\widehat{B}_{p,r}^s(\mathbb{R}^n)$ implies that the function is already an element of $B_{p,r}^s(\mathbb{R}^n)$, see Triebel [80] for details. We are going to extend this result to the anisotropic case in the following lemma.

Lemma 5.1.7. *Let* $s \in \mathbb{R}$ *with* $s > 0$. *The norms* $\|\cdot\|_{L_{\vec{p}}(\mathbb{R}^n)} + \|\cdot\|_{\widehat{B}_{\vec{p},r}^{s,\vec{a}}(\mathbb{R}^n)}$ *and* $\|\cdot\|_{B_{\vec{p},r}^{s,\vec{a}}(\mathbb{R}^n)}$ *are equivalent norms on* $B_{\vec{p},r}^{s,\vec{a}}(\mathbb{R}^n)$ *for all* $1 \le \vec{p}, r \le \infty$.

Proof: Important for the following estimates is the identity $\varphi_j(0,\cdot) = \phi_j$ for $j \in \mathbb{N}$, see Definition 3.1.2 and (2.28). It holds

$$\|f\|_{\widehat{B}_{\vec{p},r}^{s,\vec{a}}(\mathbb{R}^n)} \le \left(\sum_{j=-\infty}^{0} 2^{jsr} \|\mathscr{F}^{-1}\phi_j \mathscr{F}f\|_{L_{\vec{p}}(\mathbb{R}^n)}^r \right)^{\frac{1}{r}} + \left(\sum_{j=1}^{\infty} 2^{jsr} \|\mathscr{F}^{-1}\phi_j \mathscr{F}f\|_{L_{\vec{p}}(\mathbb{R}^n)}^r \right)^{\frac{1}{r}}$$

$$\le \left(\sum_{j=-\infty}^{0} 2^{jsr} \|\mathscr{F}^{-1}\phi_j\|_{L_1(\mathbb{R}^n)} \|f\|_{L_{\vec{p}}(\mathbb{R}^n)}^r \right)^{\frac{1}{r}} + \|f\|_{B_{\vec{p},r}^{s,\vec{a}}(\mathbb{R}^n)}.$$

By (5.3) and (5.4) we have $\|\mathscr{F}^{-1}\phi_j\|_{L_1(\mathbb{R}^n)} = \|\mathscr{F}^{-1}\phi_0\|_{L_1(\mathbb{R}^n)}$ and because $s > 0$ we obtain

$$\|f\|_{\widehat{B}_{\vec{p},r}^{s,\vec{a}}(\mathbb{R}^n)} \le c(s,r)\|f\|_{L_{\vec{p}}(\mathbb{R}^n)} + \|f\|_{B_{\vec{p},r}^{s,\vec{a}}(\mathbb{R}^n)} \le 2c(s,r)\|f\|_{B_{\vec{p},r}^{s,\vec{a}}(\mathbb{R}^n)}.$$

Theorem 3.1.8 yields $\|f\|_{L_{\vec{p}}(\mathbb{R}^n)} \le \|f\|_{B_{\vec{p},1}^{0,\vec{a}}(\mathbb{R}^n)} \le \|f\|_{B_{\vec{p},r}^{s,\vec{a}}(\mathbb{R}^n)}$ and hence the first direction. The opposite direction follows from

$$\|f\|_{B_{\vec{p},r}^{s,\vec{a}}(\mathbb{R}^n)} \le \|\mathscr{F}^{-1}\varphi_0(0,\cdot)\mathscr{F}f\|_{L_{\vec{p}}(\mathbb{R}^n)} + \left(\sum_{j=1}^{\infty} 2^{jsr} \|\mathscr{F}^{-1}\varphi_j(0,\cdot)\mathscr{F}f\|_{L_{\vec{p}}(\mathbb{R}^n)}^r \right)^{\frac{1}{r}}$$

$$\lesssim \|f\|_{L_{\vec{p}}(\mathbb{R}^n)} + \left(\sum_{j=-\infty}^{\infty} 2^{jsr} \|\mathscr{F}^{-1}\phi_j \mathscr{F}f\|_{L_{\vec{p}}(\mathbb{R}^n)}^r \right)^{\frac{1}{r}} = c\|f\|_{L_{\vec{p}}(\mathbb{R}^n)} + \|f\|_{\widehat{B}_{\vec{p},r}^{s,\vec{a}}(\mathbb{R}^n)},$$

since $\varphi_j(0,\cdot) = \phi_j$ for $j \in \mathbb{N}$. □

The following lemma extends well-known results of Bochner spaces on $(0,T) \times \Omega$ to the time-periodic setting. The result was needed in Section 4.2 to improve an estimate.

Lemma 5.1.8. *Let $\Omega \subset \mathbb{R}^n$ be a domain. For $1 < p, q < \infty$ and $k, l \in \mathbb{N}_0$ the spaces $L_{q,p}(\Omega_\mathbb{T})$ and $W_{q,p}^{k,l}(\Omega_\mathbb{T})$ are reflexive. If Ω is bounded and $l, k \in \mathbb{N}$ the embedding $W_{q,p}^{k,l}(\Omega_\mathbb{T}) \hookrightarrow W_{q,p}^{k-1,l-1}(\Omega_\mathbb{T})$ is compact.*

Proof: By standard theory it is well-known that $L_{q,p}(\mathbb{T} \times \Omega) = L_q(\mathbb{T}, L_p(\Omega))$ is reflexive since $L_p(\Omega)$ is. The space $W_{q,p}^{k,l}(\Omega_\mathbb{T})$ can be seen as a closed subspace of $L_{q,p}(\Omega_\mathbb{T})^m$ for some $m \in \mathbb{N}_0$, for details see Adams and Fournier [1, Chapter 3], and hence reflexivity follows from standard theory. The trivial embedding into $W_{q,p}^{k,l}(\Omega_\mathbb{T}) \hookrightarrow W_{q,p}^{k,l}((0, \mathcal{T}) \times \Omega)$ implies compactness, because it holds in this case and the norms coincide. $\qquad \square$

5.2 Lifting of the Divergence

We extend the results of and Farwig and Sohr [33] of solving a divergence problem with right-hand sides depending on time $t \in \mathbb{T}$. We follow the ideas of Maekawa and Sauer [66], but give additional details and explicitly investigate the time regularity of the solution. Let $\mathcal{O}$ be either $\mathbb{R}^n$ or $\mathbb{R}_+^n$, recall $\mathcal{O}_\mathbb{T} = \mathbb{T} \times \mathcal{O}$ and omit boundary conditions for $\mathbb{R}^n$. Before we start, we need to introduce the space $\widehat{W}_p^{-1}(\mathcal{O})$ for $1 < p < \infty$. It is given as the dual of $\widehat{W}_p^1(\mathcal{O})$ and for $v \in \widehat{W}_p^{-1}(\mathcal{O})$ the norm is given by

$$\|v\|_{\widehat{W}_p^{-1}(\mathcal{O})} := \sup_{\substack{\varphi \in \mathcal{D}(\overline{\mathcal{O}}) \\ \varphi \neq 0}} \frac{\langle v, \varphi \rangle}{\|\nabla \varphi\|_{L_p(\mathcal{O})}}.$$

For more details see Farwig and Sohr [33, Chapter 1]. Note that these spaces allow for a trivial extension from $\mathbb{R}_+^n$ to $\mathbb{R}^n$.

Lemma 5.2.1. *Let $1 < q, p < \infty$, $h \in W_{q,p}^{0,1}(\mathcal{O}_\mathbb{T})$ and $h \in W_q^1(\mathbb{T}, \widehat{W}_p^{-1}(\mathcal{O}))$. Then there exists a unique solution $\nabla v \in W_{q,p}^{1,2}(\mathcal{O}_\mathbb{T})$ to*

(5.5)
$$\begin{cases} \Delta v = h & \text{in } \mathcal{O}_\mathbb{T}, \\ \partial_n v = 0 & \text{on } \partial \mathcal{O}_\mathbb{T}. \end{cases}$$

Furthermore, there exists a constant $c > 0$ such that

(5.6)
$$\|\nabla v\|_{W_{q,p}^{1,2}(\mathcal{O}_\mathbb{T})} \leq c \Big(\|h\|_{W_{q,p}^{0,1}(\mathcal{O}_\mathbb{T})} + \|h\|_{W_q^1(\mathbb{T}, \widehat{W}_p^{-1}(\mathcal{O}))} \Big).$$

Additionally, if h is purely periodic so is ∇v, i.e., $\nabla v \in \mathcal{P}_\perp W_{q,p}^{1,2}(\mathcal{O}_\mathbb{T})$.

Proof: We start with the whole space $\mathbb{R}^n$. For almost all $t \in \mathbb{T}$ the function $h(t, \cdot)$ is an element of $W_p^1(\mathbb{R}^n) \cap \widehat{W}_p^{-1}(\mathbb{R}^n)$ and hence by Farwig and Sohr [33, Section 2] there exists a unique weak solution $\nabla v(t, \cdot) \in L_p(\mathbb{R}^n)$ to

$$\int_{\mathbb{R}^n} \nabla v(t, x) \nabla \varphi(x) \, \mathrm{d}x = \int_{\mathbb{R}^n} h(t, x) \varphi(x) \, \mathrm{d}x$$

for all $\varphi \in C_0^\infty(\mathbb{R}^n)$. This solution additionally satisfies

(5.7)
$$\int_{G_n} \nabla v(t, x) \nabla \psi(t, x) \, \mathrm{d}(t, x) = \int_{G_n} h(t, x) \psi(t, x) \, \mathrm{d}(t, x)$$

for all $\psi \in C_0^\infty(G_n) = C_0^\infty(\mathbb{T} \times \mathbb{R}^n)$. By taking $\psi(t,x) = \psi_1(t)\psi_2(x)$ with $\psi_1 \in C^\infty(\mathbb{T})$ and $\psi_2 \in C_0^\infty(\mathbb{R}^n)$ we see that (5.7) has a unique solution $\nabla v \in L_{q,p}(G_n)$ for every $h \in L_q(\mathbb{T}, \widehat{W}_p^{-1}(\mathbb{R}^n))$. The additional regularity of h with respect to space yields $\nabla v(t, \cdot) \in W_p^2(\mathbb{R}^n)$ by Farwig and Sohr [33]. Integrating the estimate of Farwig and Sohr over $\mathbb{T}$ yields

$$\|\nabla v\|_{W_{q,p}^{0,2}(G_n)} \lesssim \|h\|_{W_{q,p}^{0,1}(G_n)} + \|h\|_{L_q(\mathbb{T}, \widehat{W}_p^{-1}(\mathbb{R}^n))}.$$

Since $\partial_t h(t, \cdot) \in \widehat{W}_p^{-1}(\mathbb{R}^n)$ for almost all $t \in \mathbb{T}$, we obtain a unique solution $\nabla u(t, \cdot) \in L_p(\mathbb{R}^n)$ that satisfies (5.7). For $\psi \in C_0^\infty(G_n)$ it holds

$$\langle \partial_t \nabla v, \nabla \psi \rangle = -\langle \nabla v, \nabla \partial_t \psi \rangle = -\langle h, \partial_t \psi \rangle = \langle \partial_t h, \psi \rangle.$$

Because the solution is unique, we derive $\partial_t \nabla v = \nabla u \in L_{q,p}(\mathbb{R}^n)$. This implies (5.6) because $\partial_t \nabla v$ satisfies the estimate

$$\|\partial_t \nabla v\|_{L_{q,p}(G_n)} = \|\nabla u\|_{L_{q,p}(G_n)} \lesssim \|\partial_t h\|_{L_q(\mathbb{T}, \widehat{W}_p^{-1}(\mathbb{R}^n))}$$

by integrating the corresponding estimate of Farwig and Sohr.

In the half space we extend the given element h by an even extension to the whole space, $i.e.$, $Eh(t, x', x_n) := h(t, x', -x_n)$ for $x_n < 0$. Since this extension adopts the regularity of h we obtain a unique solution in the whole space. The restriction to the half space solves (5.5) and it is well-known that it satisfies the boundary data, see for example Farwig and Sohr [33, Remark 2.1] or the proof of Lemma 4.2.2. Because the construction of the solution leaves the time variable invariant, the purely periodicity of ∇v follows directly from h. $\qquad\square$

We will give a short corollary for conditions in which Lemma 5.2.1 is applicable to $\mathcal{O} = \mathbb{R}_+^n$.

Corollary 5.2.2. *Let $G \in W_{q,p}^{1,2}(\mathbb{T} \times \mathbb{R}_+^n)$ such that $G(t, x', 0) \cdot n = 0$ for all $(t, x') \in G_{n-1}$ and n denoting the outer normal vector to $\mathbb{R}_+^n$. Then Lemma 5.2.1 is applicable to $h = \operatorname{div} G$. Furthermore, it holds*

$$(5.8) \qquad \|h\|_{W_{q,p}^{0,1}(\mathbb{T} \times \mathbb{R}_+^n)} + \|h\|_{W_q^1(\mathbb{T}, \widehat{W}_p^{-1}(\mathbb{R}_+^n))} \lesssim \|G\|_{W_{q,p}^{1,2}(\mathbb{T} \times \mathbb{R}_+^n)}.$$

Proof: Let $\varphi \in C_0^\infty(\mathbb{T} \times \overline{\mathbb{R}_+^n})$, we have

$$\int\limits_{\mathbb{T} \times \mathbb{R}_+^n} \operatorname{div}(G)(t,x)\varphi(t,x) \, dt \, dx = -\int\limits_{\mathbb{T} \times \mathbb{R}_+^n} G(t,x)\nabla\varphi(t,x) \, dt \, dx + \int\limits_{\mathbb{T} \times \mathbb{R}^{n-1}} \varphi(t,x',0)G(t,x',0) \cdot n \, dt \, dx'.$$

Hence the estimate follows for $\operatorname{div}(G)$, because the boundary integral vanishes. Repeating the previous calculations with $\partial_t G$ instead of G yields the second result, because the time derivative remains zero on the boundary. $\qquad\square$

Bibliography

[1] R. A. Adams and J. J. F. Fournier. *Sobolev Spaces*, volume 140 of *Pure and Applied Mathematics (Amsterdam)*. Elsevier/Academic Press, Amsterdam, second edition, 2003.

[2] T. Akiyama. On the existence of L^p solutions of the magnetohydrodynamic equations in a bounded domain. *Nonlinear Anal.*, 54(6):1165–1174, 2003.

[3] H. Al Baba, C. Amrouche, and M. Escobedo. Analyticity of the semi-group generated by the Stokes operator with Navier-type boundary conditions on L_p-spaces. In *Recent advances in partial differential equations and applications*, volume 666 of *Contemp. Math.*, pages 23–40. Amer. Math. Soc., Providence, RI, 2016.

[4] H. Al Baba, C. Amrouche, and M. Escobedo. Semi-Group Theory for the Stokes Operator with Navier-Type Boundary Conditions on L^p-Spaces. *Arch. Ration. Mech. Anal.*, 223(2):881–940, 2017.

[5] H. Amann. Nonhomogeneous Linear and Quasilinear Elliptic and Parabolic Boundary Value Problems. In *Function spaces, differential operators and nonlinear analysis (Friedrichroda, 1992)*, volume 133 of *Teubner-Texte Math.*, pages 9–126. Teubner, Stuttgart, 1993.

[6] C. Amrouche and N. Seloula. On the Stokes equations with the Navier-type boundary conditions. *Differ. Equ. Appl.*, 3(4):581–607, 2011.

[7] C. Amrouche and N. Seloula. L^p-theory for vector potentials and Sobolev's inequalities for vector fields: application to the Stokes equations with pressure boundary conditions. *Math. Models Methods Appl. Sci.*, 23(1):37–92, 2013.

[8] W. Arendt, C. J. K. Batty, M. Hieber, and F. Neubrander. *Vector-valued Laplace transforms and Cauchy problems*, volume 96 of *Monographs in Mathematics*. Birkhäuser Verlag, Basel, 2001.

[9] W. Arendt and S. Bu. The operator-valued Marcinkiewicz multiplier theorem and maximal regularity. *Math. Z.*, 240(2):311–343, 2002.

[10] R. J. Bagby. An extended inequality for the maximal function. *Proc. Amer. Math. Soc.*, 48:419–422, 1975.

[11] H. Bahouri, J.-Y. Chemin, and R. Danchin. *Fourier Analysis and Nonlinear Partial Differential Equations*, volume 343 of *Grundlehren der Mathematischen Wissenschaften [Fundamental Principles of Mathematical Sciences]*. Springer, Heidelberg, 2011.

[12] A. Benedek, A.-P. Calderón, and R. Panzone. Convolution operators on Banach space valued functions. *Proc. Nat. Acad. Sci. U.S.A.*, 48:356–365, 1962.

[13] A. Benedek and R. Panzone. The space L^p, with mixed norm. *Duke Math. J.*, 28:301–324, 1961.

[14] M. Z. Berkolaĭko. Traces of generalized spaces of differentiable functions with mixed norm. *Dokl. Akad. Nauk SSSR*, 277(2):270–274, 1984.

[15] M. Z. Berkolaĭko. Theorems on traces on coordinate subspaces for some spaces of differentiable functions with anisotropic mixed norm. *Dokl. Akad. Nauk SSSR*, 282(5):1042–1046, 1985.

[16] M. Z. Berkolaĭko. Traces of functions in generalized Sobolev spaces with a mixed norm on an arbitrary coordinate subspace. I. *Trudy Inst. Mat. (Novosibirsk)*, 7(Issled. Geom. Mat. Anal.):30–44, 199, 1987.

[17] M. Z. Berkolaĭko. Traces of functions in generalized Sobolev spaces with a mixed norm on an arbitrary coordinate subspace. II. *Trudy Inst. Mat. (Novosibirsk)*, 9(Issled. Geom. "v tselom" i Mat. Anal.):34–41, 206, 1987.

[18] H. Brezis and P. Mironescu. Gagliardo-Nirenberg, composition and products in fractional Sobolev spaces. *J. Evol. Equ.*, 1(4):387–404, 2001.

[19] F. E. Browder. Existence of periodic solutions for nonlinear equations of evolution. *Proc. Nat. Acad. Sci. U.S.A.*, 53:1100–1103, 1965.

[20] A. Celik and M. Kyed. Nonlinear wave equation with damping: Periodic forcing and non-resonant solutions to the Kuznetsov equation. *ZAMM Z. Angew. Math. Mech.*, 98(3):412–430, 2018.

[21] A. Celik and M. Kyed. Nonlinear Acoustics: Blackstock-Crighton Equations with a Periodic Forcing Term. *J. Math. Fluid Mech.*, 21(3):Paper No. 45, 12, 2019.

[22] R. R. Coifman and G. Weiss. *Analyse harmonique non-commutative sur certains espaces homogènes*. Lecture Notes in Mathematics, Vol. 242. Springer-Verlag, Berlin-New York, 1971. Étude de certaines intégrales singulières.

[23] R. R. Coifman and G. Weiss. *Transference methods in analysis*. American Mathematical Society, Providence, R.I., 1976. Conference Board of the Mathematical Sciences Regional Conference Series in Mathematics, No. 31.

[24] R. R. Coifman and G. Weiss. Extensions of Hardy spaces and their use in analysis. *Bull. Amer. Math. Soc.*, 83(4):569–645, 1977.

[25] P. A. Davidson. *An Introduction to Magnetohydrodynamics*. Cambridge Texts in Applied Mathematics. Cambridge University Press, 2001.

[26] K. de Leeuw. On L_p Multipliers. *Ann. of Math. (2)*, 81:364–379, 1965.

[27] R. Denk, M. Hieber, and J. Prüss. Optimal L^p-L^q-estimates for parabolic boundary value problems with inhomogeneous data. *Math. Z.*, 257(1):193–224, 2007.

[28] R. Denk and M. Kaip. *General parabolic mixed order systems in L_p and applications*, volume 239 of *Operator Theory: Advances and Applications*. Birkhäuser/Springer, Cham, 2013.

[29] R. E. Edwards and G. I. Gaudry. *Littlewood-Paley and Multiplier Theory*. Springer-Verlag, Berlin-New York, 1977. Ergebnisse der Mathematik und ihrer Grenzgebiete, Band 90.

[30] T. Eiter and M. Kyed. Time-periodic linearized Navier-Stokes equations: an approach based on Fourier multipliers. In *Particles in flows*, Adv. Math. Fluid Mech., pages 77–137. Birkhäuser/Springer, Cham, 2017.

[31] K.-J. Engel and R. Nagel. *One-Parameter Semigroups for Linear Evolution Equations*, volume 194 of *Graduate Texts in Mathematics*. Springer-Verlag, New York, 2000.

[32] E. B. Fabes and N. M. Rivière. Singular integrals with mixed homogeneity. *Studia Math.*, 27:19–38, 1966.

[33] R. Farwig and H. Sohr. Generalized resolvent estimates for the Stokes system in bounded and unbounded domains. *J. Math. Soc. Japan*, 46(4):607–643, 1994.

[34] G. B. Folland. *A course in abstract harmonic analysis*. Textbooks in Mathematics. CRC Press, Boca Raton, FL, second edition, 2016.

[35] G. P. Galdi. *An introduction to the mathematical theory of the Navier-Stokes equations*. Springer Monographs in Mathematics. Springer, New York, second edition, 2011. Steady-state problems.

[36] G. P. Galdi. Existence and uniqueness of time-periodic solutions to the Navier-Stokes equations in the whole plane. *Discrete Contin. Dyn. Syst. Ser. S*, 6(5):1237–1257, 2013.

[37] G. P. Galdi. On Time-Periodic Flow of a Viscous Liquid past a Moving Cylinder. *Arch. Ration. Mech. Anal.*, 210(2):451–498, 2013.

[38] G. P. Galdi and M. Kyed. Time-periodic flow of a viscous liquid past a body. In *Partial differential equations in fluid mechanics*, volume 452 of *London Math. Soc. Lecture Note Ser.*, pages 20–49. Cambridge Univ. Press, Cambridge, 2018.

[39] A. G. Georgiadis and M. Nielsen. Pseudodifferential operators on mixed-norm Besov and Triebel-Lizorkin spaces. *Math. Nachr.*, 289(16):2019–2036, 2016.

[40] D. Gilbarg and N. S. Trudinger. *Elliptic Partial Differential Equations of Second Order*, volume 224 of *Grundlehren der Mathematischen Wissenschaften [Fundamental Principles of Mathematical Sciences]*. Springer-Verlag, Berlin, second edition, 1983.

[41] L. Grafakos. *Classical Fourier Analysis*, volume 249 of *Graduate Texts in Mathematics*. Springer, New York, third edition, 2014.

[42] P. Grisvard. *Elliptic Problems in Nonsmooth Domains*, volume 24 of *Monographs and Studies in Mathematics*. Pitman (Advanced Publishing Program), Boston, MA, 1985.

[43] M. Gunzburger and C. Trenchea. Analysis and discretization of an optimal control problem for the time-periodic MHD equations. *J. Math. Anal. Appl.*, 308(2):440–466, 2005.

[44] P. Hess and T. Kato. Perturbation of closed operators and their adjoints. *Comment. Math. Helv.*, 45:524–529, 1970.

[45] M. Hieber and C. Stinner. Strong time periodic solutions to Keller-Segel systems: An approach by the quasilinear Arendt-Bu theorem. *J. Differential Equations*, 269(2):1636–1655, 2020.

[46] L. Hörmander. *The Analysis of Linear Partial Differential Operators. I.* Classics in Mathematics. Springer-Verlag, Berlin, 2003. Distribution theory and Fourier analysis, Reprint of the second (1990) edition [Springer, Berlin; MR1065993 (91m:35001a)].

[47] T. Hytönen and P. Portal. Vector-valued multiparameter singular integrals and pseudodifferential operators. *Adv. Math.*, 217(2):519–536, 2008.

[48] S. Ibrahim, P. G. Lemarié-Rieusset, and N. Masmoudi. Time-periodic forcing and asymptotic stability for the Navier-Stokes-Maxwell equations. *Comm. Pure Appl. Math.*, 71(1):51–89, 2018.

[49] J. D. Jackson. *Classical Electrodynamics*. John Wiley & Sons, Inc., New York-London-Sydney, 2. ed. edition, 1975.

[50] B. Jessen, J. Marcinkiewicz, and A. Zygmund. Note on the differentiability of multiple integrals. *Fundamenta Mathematicae*, 25(1):217–234, 1935.

[51] J. Johnsen. Pointwise multiplication of Besov and Triebel-Lizorkin spaces. *Math. Nachr.*, 175:85–133, 1995.

[52] J. Johnsen. Elliptic boundary problems and the Boutet de Monvel calculus in Besov and Triebel-Lizorkin spaces. *Math. Scand.*, 79(1):25–85, 1996.

[53] J. Johnsen, S. Munch Hansen, and W. Sickel. Anisotropic, mixed-norm Lizorkin-Triebel spaces and diffeomorphic maps. *J. Funct. Spaces*, 2014.

[54] J. Johnsen, S. Munch Hansen, and W. Sickel. Anisotropic Lizorkin-Triebel spaces with mixed norms—traces on smooth boundaries. *Math. Nachr.*, 288(11-12):1327–1359, 2015.

[55] J. Johnsen and W. Sickel. A direct proof of Sobolev embeddings for quasi-homogeneous Lizorkin-Triebel spaces with mixed norms. *J. Funct. Spaces Appl.*, 5(2):183–198, 2007.

[56] J. Johnsen and W. Sickel. On the trace problem for Lizorkin-Triebel spaces with mixed norms. *Math. Nachr.*, 281(5):669–696, 2008.

[57] T. Kato. *Perturbation theory for linear operators.* Classics in Mathematics. Springer-Verlag, Berlin, 1995. Reprint of the 1980 edition.

[58] P. Krée. Propriétés de continuité dans L^p de certains noyaux. *Boll. Un. Mat. Ital. (3)*, 22:330–344, 1967.

[59] D. S. Kurtz. Classical operators on mixed-normed spaces with product weights. *Rocky Mountain J. Math.*, 37(1):269–283, 2007.

[60] M. Kyed. Time-Periodic Solutions to the Navier-Stokes Equations. *Habilitationsschrift, Technische Universität Darmstadt*, 2012.

[61] M. Kyed and J. Sauer. A method for obtaining time-periodic L^p estimates. *J. Differential Equations*, 262(1):633–652, 2017.

[62] Ladyženskaya O. A. and Solonnikov V. A. The linearization principle and invariant manifolds for problems of magnetohydrodynamics. *Journal of Soviet Mathematics*, 8(4):384–422, 1977.

[63] L. D. Landau and E. M. Lifshitz. *Electrodynamics of continuous media.* Course of Theoretical Physics, Vol. 8. Translated from the Russian by J. B. Sykes and J. S. Bell. Pergamon Press, Oxford-London-New York-Paris; Addison-Wesley Publishing Co., Inc., Reading, Mass., 1960.

[64] X. Li and D. Wang. Global solutions to the incompressible magnetohydrodynamic equations. *Commun. Pure Appl. Anal.*, 11(2):763–783, 2012.

[65] P. I. Lizorkin. Multipliers of Fourier Integrals and Bounds of Convolutions in Spaces with Mixed Norm. Applications. *Izv. Akad. Nauk SSSR Ser. Mat.*, 34:218–247, 1970.

[66] Y. Maekawa and J. Sauer. Maximal Regularity of the Time-Periodic Stokes operator on Unbounded and Bounded Domains. *J. Math. Soc. Japan*, 69(4):1403–1429, 2017.

[67] J. Marschall. Nonregular pseudo-differential operators. *Z. Anal. Anwendungen*, 15(1):109–148, 1996.

[68] E. A. Notte, M. D. Rojas, and M. A. Rojas. Periodic strong solutions of the magnetohydrodynamic type equations. *Proyecciones*, 21(3):199–224, 2002.

[69] P. H. Rabinowitz. Periodic solutions of nonlinear hyperbolic partial differential equations. *Comm. Pure Appl. Math.*, 20:145–205, 1967.

[70] P. H. Rabinowitz. Periodic solutions of nonlinear hyperbolic partial differential equations. II. *Comm. Pure Appl. Math.*, 22:15–39, 1968.

[71] W. Rudin. *Fourier analysis on groups*. Interscience Tracts in Pure and Applied Mathematics, No. 12. Interscience Publishers (a division of John Wiley and Sons), New York-London, 1962.

[72] T. Runst and W. Sickel. *Sobolev spaces of fractional order, Nemytskij operators, and nonlinear partial differential equations*, volume 3 of *De Gruyter Series in Nonlinear Analysis and Applications*. Walter de Gruyter & Co., Berlin, 1996.

[73] S. Sato. Weak type estimates for functions of Marcinkiewicz type with fractional integrals of mixed homogeneity. *Math. Scand.*, 125(1):135–162, 2019.

[74] H.-J. Schmeisser and H. Triebel. *Topics in Fourier analysis and function spaces*. A Wiley-Interscience Publication. John Wiley & Sons, Ltd., Chichester, 1987.

[75] M. Sermange and R. Temam. Some Mathematical Questions Related to the MHD Equations. *Comm. Pure Appl. Math.*, 36(5):635–664, 1983.

[76] A. Seyfert. *The Helmholtz-Hodge Decomposition in Lebesgue Spaces on Exterior Domains and Evolution Equations on the Whole Real Time Axis*. PhD thesis, Technische Universität Darmstadt, 2018.

[77] E. M. Stein. *Singular Integrals and Differentiability Properties of Functions*. Princeton Mathematical Series, No. 30. Princeton University Press, Princeton, N.J., 1970.

[78] E. M. Stein. *Harmonic Analysis: Real-Variable Methods, Orthogonality, and Oscillatory Integrals*, volume 43 of *Princeton Mathematical Series*. Princeton University Press, Princeton, NJ, 1993. With the assistance of Timothy S. Murphy, Monographs in Harmonic Analysis, III.

[79] H. Triebel. *Interpolation theory, function spaces, differential operators*, volume 18 of *North-Holland Mathematical Library*. North-Holland Publishing Co., Amsterdam-New York, 1978.

[80] H. Triebel. *Theory of function spaces*, volume 78 of *Monographs in Mathematics*. Birkhäuser Verlag, Basel, 1983.

[81] P. Weidemaier. Existence results in L_p-L_q spaces for second order parabolic equations with inhomogeneous Dirichlet boundary conditions. In *Progress in partial differential equations, Vol. 2 (Pont-à-Mousson, 1997)*, volume 384 of *Pitman Res. Notes Math. Ser.*, pages 189–200. Longman, Harlow, 1998.

[82] P. Weidemaier. Maximal regularity for parabolic equations with inhomogeneous boundary conditions in Sobolev spaces with mixed L_p-norm. *Electron. Res. Announc. Amer. Math. Soc.*, 8:47–51, 2002.

[83] P. Weidemaier. Vector-valued Lizorkin-Triebel spaces and sharp trace theory for fucntions in Sobolev spaces with mixed L_p-norm in parabolic problems. *Mat. Sb.*, 196(6):3–16, 2005.

[84] N. Yamaguchi. On an existence theorem of global strong solutions to the magnetohydrodynamic system in three-dimensional exterior domains. *Differential Integral Equations*, 19(8):919–944, 2006.

[85] M. Yamazaki. A quasihomogeneous version of paradifferential operators. I. Boundedness on spaces of Besov type. *J. Fac. Sci. Univ. Tokyo Sect. IA Math.*, 33(1):131–174, 1986.

[86] M. Yamazaki. A quasihomogeneous version of paradifferential operators. II. A symbol calculus. *J. Fac. Sci. Univ. Tokyo Sect. IA Math.*, 33(2):311–345, 1986.

[87] Z. Yoshida and Y. Giga. On the Ohm-Navier-Stokes system in magnetohydrodynamics. *J. Math. Phys.*, 24(12):2860–2864, 1983.

[88] K. Yosida. *Functional Analysis*. Classics in Mathematics. Springer-Verlag, Berlin, 1995. Reprint of the sixth (1980) edition.

Curriculum Vitae

17/09/87	Born in Braunschweig, Germany
10/08 – 03/13	**Bachelor's studies**, *Technische Universität Darmstadt*, Darmstadt, Germany, Bachelor of Science Mathematics Bachelor's thesis: *H-Maße und ihre Anwendungen*
04/13 – 03/15	**Master's studies**, *Technische Universität Darmstadt*, Darmstadt, Germany, Master of Science Mathematics Master's thesis: *Multiplier Estimates in Spaces with Mixed Norms*
04/15 – 09/20	**Doctoral studies**, *Technische Universität Darmstadt*, Darmstadt, Germany, research assistant in the working group *Partielle Differentialgleichungen*
07/07/20	**Submission of the doctor's thesis (Dissertation)** *Time-Periodic Solutions to the Equations of Magnetohydrodynamics with Background Magnetic Field* at *Technische Universität Darmstadt*, Darmstadt, Germany
08/14/20	**Defense of the doctor's thesis**